MW01099063

Solar Hydrogen Generation

Solar Hydrogen Generation

Transition Metal Oxides in Water Photoelectrolysis

Jinghua Guo

Xiaobo Chen

New York Chicago San Francisco
Lisbon London Madrid Mexico City
Milan New Delhi San Juan
Seoul Singapore Sydney Toronto

The McGraw·Hill Companies

Cataloging-in-Publication Data is on file with the Library of Congress

Solar Hydrogen Generation: Transition Metal Oxides in Water Photoelectrolysis

1 2 3 4 5 6 7 8 9 0 DOC/DOC 1 9 8 7 6 5 4 3 2 1

ISBN: 978-0-07-170126-6
MHID: 0-07-170126-5

The pages within this book were printed on acid-free paper.

Sponsoring Editor
Michael Penn

Acquisitions Coordinator
Bridget Thoreson

Editorial Supervisor
David E. Fogarty

Project Manager
Neha Rathor

Copy Editor
Lisa McCoy

Proofreader
Manisha Sinha

Indexer
Robert Swanson

Production Supervisor
Pamela A. Pelton

Composition
Neuetype

Art Director, Cover
Jeff Weeks

While sitting in the traffic jam again to get through the Caldecott Tunnel on the way to San Francisco and watching aimlessly the ongoing Fourth Bore construction project, Jessica, my daughter, asked if it would be possible to have a flying car and what would be a clean fuel to use. While I wondered if it might take thirty years to see hydrogen-fueled flying cars and hydrogen harvested from sunlight-assistant water splitting, she jumped to her next question—why does the sun shine and for how long? Well, it burns hydrogen too, in a somewhat different way, and will continue to do so for probably another five billions years!

About the Authors

Jinghua Guo, Ph.D., is a staff scientist of Advanced Light Source at Lawrence Berkeley National Laboratory. He was previously a faculty member at Uppsala University, Sweden. Dr. Guo is the author of more than 200 peer-reviewed scientific publications, an editor of the *International Journal of Nanotechnology,* guest editor for the *Journal of Electron Spectroscopy and Related Phenomena,* and a reviewer for scientific journals, including *Physical Review Letters, Nature Chemistry, Nano Letters,* and *Journal of the American Chemical Society.*

Xiaobo Chen, Ph.D., is an assistant professor in the Department of Chemistry at the University of Missouri–Kansas City. He was previously a research scientist at Lawrence Berkeley National Laboratory and University of California–Berkeley. Dr. Chen has published 40 peer-reviewed scientific articles with more than 6,000 citations, holds three U.S. and international patents, is the chair for Materials Research Society Spring Meeting: Titanium Dioxide Nanomaterials in 2011 and 2012, and is a reviewer for many scientific journals, including *Science, Journal of the American Chemical Society,* and *Advanced Materials.*

Contents

Preface

This book is intended for researchers and graduate students of materials science and physical chemistry as an extended review in renewable energy science research. In addition, it provides some counts of the state-of-art research in hydrogen generation from photosynthesis.

The search for new ways to efficiently produce clean energy has been rapidly becoming one of the most pressing technological challenges that we are facing. At the same time, enormous progress has been made in developing new materials that are tailored by nanostructuring, supra-molecular assembly, and many other new synthesis methods. Such novel materials might be the key to producing renewable energy through efficient solar energy conversion.

Growing energy consumption worldwide, combined with global warming due to burning of fossil fuels, necessitates the discovery of new materials for novel highly efficient solar energy conversion and storage. This is also an opportunity for condensed matter physics/chemistry and bioscience in fundamental science and applied technology. It becomes increasingly important to understand the fundamental scientific problems with discovering new efficient materials for solar energy conversion and storage applications.

Solar energy conversion requires materials that simultaneously fulfill several requirements regarding their electronic structure and chemical properties in order to ensure stability and efficient utilization of the solar spectrum. No efficient material has been established today that would fulfill all of the requirements; hence, there is an urgent need to understand and tailor materials that will meet the desired requirements. The existing photocatalytic materials, for example, currently do not make use of the entire solar spectrum, but the current class of candidate materials that promise a better spectral utilization have other fundamental shortcomings that limit their overall achievable efficiency. Tandem cells could be a solution, but none of the low-cost materials and preparation processes have produced highly efficient candidates. It is clear that many materials breakthroughs are required to diversify the energy portfolio of the world.

The inherent complexity associated with the materials needed for energy technologies remains a challenge both for experimentalists and theorists, and it becomes a central mission in the first half of twenty-first century. How can the required science and technology breakthroughs be achieved? It is increasingly accepted that the design and optimization of novel materials are strongly enhanced by a cycle. Such a cycle begins with an increased and detailed fundamental knowledge of the status quo materials, an insight-based establishment of a set of desirable parameters, the skills of material scientists to produce such materials, and finally the availability of experimental characterization and theoretical interpretation tools that allow an atomistic comparison between the parameters of the materials produced and the originally envisioned set of parameters defined as desirable. These results and discovered materials then serve as the status quo, and the cycle begins anew.

The ever-increasing demand to diversify the energy portfolio, in particular to minimize environmental impact while supplying the increasing global energy needs, has intensified the urgency to develop alternative energy sources and carriers. Significant research efforts are under way in a broad range of fundamental sciences and applied technologies that have evolved to develop a consensus. Areas of fundamental research to be addressed include the following:

1. Photovoltaics for converting sunlight to electricity
2. Artificial photosynthesis for converting sunlight to fuel
3. Energy efficiency with designed catalysts
4. Carbon capture and carbon chemistry
5. Energy storage and batteries

These are profound challenges, and scientists need to develop advanced theoretical methods and sharper experimental tools that are capable of providing in-depth understanding and ultimate control of these processes to achieve ultimate goals. This book touches on only one of these scientific research areas: photosynthesis for water splitting.

How does one learn about the electronic properties of novel materials for water splitting to determine the efficient catalyst activity and selectivity? About the electron-hole pair formation at the interface of water and catalyst surface upon absorption of sunlight? About bandgaps, band levels, and the band structure of semiconducting catalysts that are of crucial importance in photoelectrochemical and photocatalytic applications? About generation of chemical fuel from direct photooxidation of water under sunlight without external bias? These are profound challenges that have been with us for decades, but are more important than ever today. The intellectual merit of the scientific research will allow scientists to address these questions through the use of theoretical modeling and various experimental tools, including synchrotron radiation based x-ray spectroscopic and microscopic tools fully optimized for the research of renewable energy science and technology.

So far, most of the developments in renewable energy materials have been achieved by the well-tested, Edisonian method of trial and error. However, progress has been slow when compared to the rapid advances in other electronic devices, exemplified by Moore's law for device density in silicon microchips or the increase in magnetic data storage density. This situation calls for a new strategy, where slow evolution of traditional concepts is accelerated by feedback from spectroscopy. A close feedback loop between synthesis, characterization, and theoretical prediction enables a more rational design of new materials than does trial and error. For example, by tailoring the electronic energy levels of the absorber molecules or semiconductors, the donor/acceptor for electron-hole separation, and the eventual transport to a conducting electrode, one can minimize the energy loss in a solar cell. These energy levels can be determined by incisive, element-specific spectroscopic techniques based on synchrotron radiation x-ray spectroscopic and microscopic tools.

This book has been organized in the following way. Chapter 1 describes the basic processes of electrochemistry and photocatalysis for hydrogen production. In an electrochemical reaction, water can be decomposed into hydrogen (H_2) and oxygen (O_2) gases when an electric current is passed through the water. Conversely, photocatalysis splits water directly into H_2 and O_2 by utilizing the sunlight energy only. Chapter 2 focuses on the basics in photocatalytic reactions, and briefly discusses the potential catalysts, including semiconductor catalysts, biomimetic catalysts, and nanostructured catalysts. There are a number of basic points to be considered for efficient photocatalytic water splitting reactions, including stoichiometry of H_2 and O_2 evolution, turnover number (TON), quantum yield, time course, and photoresponse. Over the last three decades the scientific and engineering interest in the application of semiconductor photocatalysis has grown exponentially. Chapter 3 shows semiconductor photocatalysts, with a primary focus on transition metal oxides, as a durable photocatalyst has been applied to a variety of problems of environmental interest in addition to water and air purification. Chapters 4 and 5 illustrate the corresponding crystal structure versus electronic structure and optical properties versus light absorption of transition metal oxides, respectively. Chapter 6 details the impurity and doped photocatalysts, and the integrated organic and inorganic systems. Chapter 7 briefly discusses surface and interface chemistry, and nanostructure and morphology in photocatalysis applications. In Chapter 8, we introduce the basics of soft x-ray absorption (XAS) and soft x-ray emission spectroscopy (XES), and resonant inelastic soft x-ray scattering (RIXS) followed by descriptions of instrumentation, including beamline, endstation, and spectrometer. Chemical cells are designed for in situ study of the electronic structure of samples in a gas or liquid environment. The application of XAS, XES, and RIXS on TiO_2 crystals of rutile and anatase phases has yielded characteristic fingerprints that provide information on geometric structure, bandgap, and doping effects. A number of in situ electronic structure studies are also presented by way of example.

I would like to express my gratitude to the many colleagues who collaborated over the years on the topics referred to in this book and who helped me in the various stages, to name a few: August Augustsson, Sergei Butorin, Chinglin Chang, Jen-Lung Chen, Jau-Wen Chiou, Xingyi Deng, Frank de Groot, Chungli Dong, Laurent Duda, Heinz Frei, Per-Anders Glans, Xiao Cheng, Peng Jiang, Feng Jiao, Stepan Kashtanov, Yi-Sheng Liu, Yi Luo, Cormac McGuinness, Joseph Nordgren, Miquel Salmeron, Vittal Yachandra, Junko Yano, Kevin Smith, Lionel Vayssieres, Gunnar

Westin, Samuel Mao, and Hans Ågren. I gratefully acknowledge the cooperation of the publisher, especially Neha Rathor, at Neuetype and Taisuke Soda, Michael Mulcahy, and Michael Penn at McGraw Hill. Finally, I would like to thank my family, Hui Cheng, Jessica, and Nicole, for their understanding and support.

JINGHUA GUO

Berkeley, California

Solar Hydrogen Generation

CHAPTER 1

Hydrogen Generation: Electrochemistry and Photoelectrolysis

Sunlight can provide a very significant contribution to our future energy use, if efficient and inexpensive systems using earth-abundant materials could be devised for the energy conversion. Solar energy, for example, can be converted to electricity and chemical fuels for energy use and storage (see Figure 1.1).[1–4] Such a direct conversion of solar energy is important because it does not pollute (no acid rain or carbon dioxide emission). Moreover, there is much more solar energy available than needed, and it will be renewed as long as the sun shines.

There has been intensive research in exploring the catalysts that are capable of using sunlight to split water to generate hydrogen gas, which is a complete clean fuel. Unfortunately, the catalysts being discovered so far that work under harsh chemical conditions with high efficiency are made from platinum, a rare and expensive metal. In many of the alternate renewable energy conversion technologies, it is equally apparent that the significant breakthroughs are needed for optimized low-cost materials with custom-designed properties.

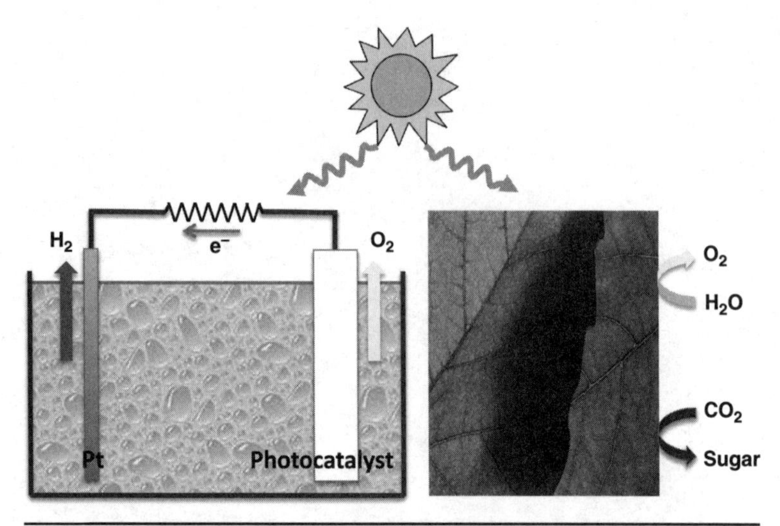

Figure 1.1 Energy-conversion strategies for creating chemical fuel from sunlight. Right, in photosynthesis, plants use solar radiation, in conjunction with CO_2 and water, to produce sugars (the fuel) and O_2. Left, a system comprising a semiconducting photocatalyst and metal (Pt) electrode immersed in water produces H_2 as the potential fuel. Under sunlight irradiation, photoexcited electrons reduce water to give H_2, whereas the electron vacancies oxidize water to give O_2.

There are fundamental requirements and still basic scientific challenges for the photocatalysts that convert solar energy:

1. Efficient absorption of sunlight covering a large fraction of the solar spectrum by photocatalysts with suitable bandgaps.
2. Electrons and holes created upon absorption of sunlight are sufficiently spatially separated to prevent recombination as well as suffer minimal loss of energy due to scattering.

To meet these challenges, new advanced tools must be brought to bear on materials that can exhibit the novel properties required under a real-world environment. While utilizing solar energy will require new materials developed through the control of atomic, chemical, and electronic structure, achieving such control requires an intimate collaboration between materials synthesis and characterization of the electronic properties of complex materials.

The conversion of solar energy to chemical energy through suitable photochemical reactions was recognized nearly a century ago. The plants hold the secret by creating new compounds to master the photochemical processes.[5–6] The adenosine triphosphate (ATP: $C_{10}H_{16}N_5O_{13}P_3$) is a nucleotide that contains a large amount of chemical energy stored in its high-energy phosphate bonds. It releases energy when it is broken down into adenosine diphosphate (ADP), which can be used for many metabolic processes. Similar to the role of the phosphate bond in ATP in biology, the H-H bond in H_2 can become a chemical energy for carbon-free energy utilization if H_2 can be readily generated from its stable carrier, water. The research is focused on using the solar energy to drive the thermodynamically uphill splitting of liquid water to produce hydrogen gas (H_2) and oxygen gas (O_2).

To give an example, an important experiment of the generation of H_2 from direct photo-oxidation of water was performed in 1970s.[7–9] To split water into H_2 and O_2 using only solar energy, the valence band of the semiconducting photocatalyst has to be located at a lower energy level than the chemical potential of oxygen evolution (H_2O/O_2), and the conduction band has to be located at a higher energy level than the chemical potential of hydrogen evolution (H_2O/H_2). It has been hard to discover a single photocatalyst that fulfills these requirements; thus, a more complex system has to be considered for the photocatalytic reactions. If the valence and conduction band energies do not fulfill this requirement, an external bias has to be applied to assist the photocatalytic process, which in turn substantially reduces the overall efficiency.

1.1 Electrochemistry

Electrolysis of water is a chemical process that splits water (H_2O) into oxygen (O_2) and hydrogen (H_2) by passing an electric current through the water. In an electrolysis setup two contacting plates serve as the electrodes (typically made from the inert metals such as platinum). By design, hydrogen gas will form on the surface of the cathode (the negatively charged electrode, where electrons appear), and oxygen will form on the surface of the anode (the positively charged electrode). For ideal faradaic efficiency, the production of hydrogen is twice the number of moles of oxygen, and both are proportional to the total electrical charge conducted by the solution. However, in many cells competing side reactions dominate, resulting in different products and less efficiency.

Electrolysis of pure water requires excess energy in the form of overpotential to overcome various activation barriers. Without the excess energy the electrolysis of pure water occurs very slowly due to the limited self-ionization of water. Pure water has an electrical conductivity about one-millionth that of seawater. The efficacy of electrolysis is increased through the addition of an electrolyte (such as a salt, an acid, or a base) and the use of electrocatalysts. Currently, the electrolytic process is rarely used in industrial applications, since hydrogen can be produced more affordably from fossil fuels.

English chemists first used electricity to split water more than 200 years ago. The reaction requires two separate catalytic steps: (i) The positively charged electrode, or anode, removes electrons from hydrogen atoms in water molecules, which results in hydrogen ions (protons) created that break away from their oxygen atoms. The anode catalyst then welds pairs of oxygen atoms together to form O_2. (ii) The free protons drift through the solution to the cathode, where pairs of protons and electrons combine to form molecular hydrogen (H_2).

In 1789, Jan Rudolph Deiman and Adriaan Paets van Troostwijk used[10–14] an electrostatic machine to produce electricity, which was discharged on gold electrodes in a Leyden jar with water.[15] In 1800, Alessandro Volta invented the voltaic pile, and then William Nicholson and Anthony Carlisle used it for the electrolysis of water.[16–17] In 1869, Zénobe Gramme invented the Gramme machine and made electrolysis of water a cheaper method to produce hydrogen gas.

In water splitting, at the anode side, an oxidation reaction occurs to form oxygen gas:

$$\text{Anode (oxidation process): } 2H_2O(l) \rightarrow O_2(g) + 4H^+(aq) + 4e^-$$

and giving electrons to the anode to complete the circuit, a reduction reaction takes place at the cathode side, with electrons (e^-) from the cathode being given to protons to form hydrogen gas (the half-reaction balanced with acid):

$$\text{Cathode (reduction process): } 2H^+(aq) + 2e^- \rightarrow H_2(g)$$

The same half-reactions can also be balanced with a base. To add half-reactions they must both be balanced with either acid or base.

$$\text{Cathode (reduction): } 2H_2O(l) + 2e^- \rightarrow H_2(g) + 2OH^-(aq)$$

$$\text{Anode (oxidation): } 4OH^-(aq) \rightarrow O_2(g) + 2H_2O(l) + 4e^-$$

Combining either half-reaction pair yields the same overall decomposition of liquid water into oxygen and hydrogen gases:

$$\text{Overall reaction: } 2H_2O(l) \rightarrow 2H_2(g) + O_2(g)$$

The number of H_2 molecules produced is thus twice the number of O_2 molecules. Assuming equal temperature and pressure for both gases, the produced H_2 gas has therefore twice the volume of the produced O_2 gas. The number of electrons pushed through the water is twice the number of generated H_2 molecules and four times the number of generated O_2 molecules.

Decomposition of pure water into hydrogen and oxygen at standard temperature and pressure is not favorable in thermodynamical terms.

$$\text{Anode (oxidation): } 2H_2O(l) \rightarrow O_2(g) + 4H^+(aq) + 4e^-;\quad E^o_{oox} = -1.23\text{ V}$$

$$\text{Cathode (reduction): } 2H^+(aq) + 2e^- \rightarrow H_2(g);\ E^o_{red} = 0.00\text{ V}$$

The standard potential of the water electrolysis cell is −1.23 V at 25°C at pH 0 ($H^+ = 1.0$ M). It is also −1.23 V at 25°C at pH 7 ($H^+ = 1.0 \times 10^{-7}$ M) based on the Nernst equation.

The negative voltage indicates the Gibbs free energy for electrolysis of water is greater than zero for these reactions. So, the reaction cannot occur without additional energy from an external electrical power source.

When the electrolysis occurs in pure water, H^+ ions (protons) will accumulate at the anode and OH^- ions will accumulate at the cathode. The water becomes acidic near the anode and basic near the cathode. Thus, the negative hydroxyl ions mostly combine with the positive hydronium ions (H_3O^+) to form water at the anode. The positive hydronium ions mostly combine with negative hydroxyl ions to form water at the cathode. Relatively few hydronium (hydroxyl) ions reach the cathode (anode), which causes a concentration overpotential at both electrodes.

Pure water is a fairly good insulator due to the low autoionization, $K_w = 10 \times 10^{-14}$ at room temperature, and thus, pure water conducts current poorly; 0.055 Scm^{-1}. A very large potential is needed to increase the autoionization of water, otherwise, the electrolysis of pure water proceeds very slowly due to the low overall conductivity.

After adding a water-soluble electrolyte, the conductivity of the water will rise considerably. The electrolyte disassociates into cations and anions; the anions move toward the anode and neutralize the buildup of positively charged H^+ there, and the cations move toward the cathode and neutralize the buildup of negatively charged OH^- there. This process allows a continued flow of electricity.[18]

There are a number of factors to consider when choosing an electrolyte, since an anion from the electrolyte is in competition with the hydroxide ions to give up an electron. An electrolyte anion with less standard electrode potential than hydroxide will be oxidized instead of the hydroxide, and no oxygen gas will be produced. A cation with a greater standard electrode potential than a hydrogen ion will be reduced instead and no hydrogen gas will be produced.

Some cations have lower electrode potential than H^+ and are therefore suitable for use as electrolyte cations, such as Li^+, Rb^+, K^+, Cs^+, Ba^{2+}, Sr^{2+}, Ca^{2+}, Na^+, and Mg^{2+}. Sodium and lithium are frequently used, as they form inexpensive, soluble salts.

When an acid is used as the electrolyte, the cation is H^+ and there is no competitor for the H^+ created by disassociating water. The most commonly used anion is sulfate (SO_4^{2-}), which is very difficult to oxidize, with the standard potential for oxidation of this ion to the peroxodisulfate ion being –2.05 V. Strong acids such as sulfuric acid (H_2SO_4) and strong bases such as potassium hydroxide (KOH) and sodium hydroxide (NaOH) are frequently used as electrolytes.

About 4 percent of hydrogen gas produced worldwide is created by electrolysis. The hydrogen produced through such electrolysis is a side product in the production of chlorine. This is a prime example of a competing side reaction.

$$2NaCl + 2H_2O \rightarrow Cl_2 + H_2 + 2NaOH$$

The electrolysis of saltwater, a water sodium chloride mixture, is only half the electrolysis of water, since the chloride ions are oxidized to chlorine rather than water being oxidized to oxygen.

The majority of industrial hydrogen is produced from fossil fuels. This has been used to make ammonia for fertilizer via the Haber process and to convert heavy petroleum sources to lighter fractions via hydrocracking. The H_2 production usually involves the formation of synthesis gas: a mixture of H_2 and CO. In this process, hydrogen can be enriched through the water gas shift reaction, where the carbon monoxide is reacted with water to produce more H_2 with CO_2 as a by-product.

Water electrolysis does not convert 100 percent of the electrical energy into the chemical energy of hydrogen. The process requires more extreme potentials than what would be expected based on the cell's total reversible reduction potentials. This excess potential accounts for various forms of overpotential by which the extra energy is eventually lost as heat. For a well-designed cell the largest overpotential occurs with the four-electron oxidation of water to oxygen at the anode. An effective electrocatalyst to facilitate this reaction has not yet been developed. Platinum alloys are the default state of the art for this oxidation. Developing a cheaper effective electrocatalyst for this reaction would be a great advance.

In 2008, a group led by Daniel Nocera announced a potentially more efficient catalyst for artificial photosynthesis composed of a cobalt metal, phosphate, and an electrode.[19] Other researchers are pursuing carbon-based catalysts. The reverse reaction, the reduction of oxygen to water, is responsible for the greatest loss of efficiency in fuel cells. The simpler two-electron reaction to produce hydrogen at the cathode can be electrocatalyzed with almost no reaction overpotential by platinum or, in theory, a hydrogenase enzyme. If other, less effective, materials are used for the cathode, then another large overpotential must be paid.

The energy efficiency of water electrolysis varies widely with the numbers cited on the optimistic side. Some report

50 to 80 percent.[20] These values refer only to the efficiency of converting electrical energy into hydrogen's chemical energy. The energy lost in generating the electricity is not included. For instance, when considering a power plant that converts the heat of nuclear reactions into hydrogen via electrolysis, the total efficiency may be closer to 30 to 45 percent.

1.2 Photocatalysis

Photocatalysis involves splitting water directly into hydrogen (H_2) and oxygen (O_2) using the energy of sunlight. Hydrogen generated from water splitting holds particular interest since it utilizes an abundant natural resource. Photocatalysts used in water splitting have several basic requirements for the bandgap and band levels. The water reduction reaction to form H_2 occurs at 0 V, so the conduction band needs to be at a potential <0 V. The water oxidation reaction to form O_2 occurs at 1.23 V, so the valence band needs to be at a potential >1.23 V. Thus, the minimum bandgap for successful water splitting is 1.23 eV, corresponding to light of 1008 nm, electrochemical requirements that can theoretically reach down into infrared light.

Honda and Fujishima reported the first demonstration of water splitting using a TiO_2 electrode under light in the early 1970s.[7] TiO_2 has a bandgap of 3.2 eV, so when irradiated with ultraviolet (UV) light, electrons and holes are generated and separated to different electrodes, as shown in Figure 1.2.[7] The photogenerated electrons reduce water to form H_2 on a Pt counter electrode, while holes oxidize water to form O_2 on the TiO_2 electrode. Since then, scientists have extensively studied water splitting to search for semiconductor photoelectrodes and photocatalysts. However, efficient photocatalytic materials for splitting water into H_2 and O_2 under sunlight have not been found for practical applications.

Photocatalysts need to meet the special band requirements and typically have dopants and/or co-catalysts added to optimize their performance. Using TiO_2 as a semiconductor with the proper band structure has been the most studied model system. However, due to the relatively positive conduction band of TiO_2, there is little driving force for H_2 production, so a co-catalyst such as Pt is used to increase the H_2

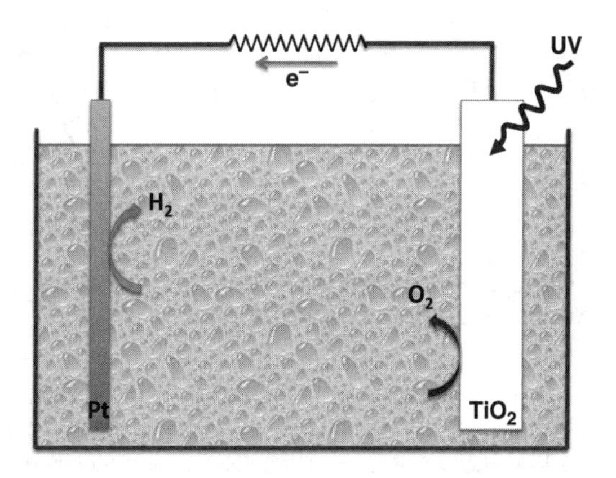

FIGURE 1.2 Honda/Fujishima effect of water splitting using a TiO_2 photoelectrode.

production. Also, in order to absorb visible light, the bandgap needs to be reduced.

In addition, the photocatalyst needs to be durable in liquid water, as it can suffer from catalyst decay and recombination under reaction conditions. Using a sulfide-based photocatalyst such as CdS can be a problem, as the sulfide in the catalyst is oxidized to elemental sulfur at the same potentials used to split water. Thus, when sulfide-based photocatalysts are used, sacrificial reagents such as sodium sulfide have to be added to replenish any sulfur lost.[21]

For efficient water splitting, photocatalysts must possess a few key elements. One key element is that H_2 and O_2 evolution should occur in a stoichiometric 2:1 ratio. The prime measure of photocatalyst effectiveness is quantum yield (QY), which is[21]:

$$\mathrm{QY}(\%) = \frac{N_{\mathrm{re}}}{N_{\mathrm{ip}}} \times 100 \tag{1.1}$$

where N_{re} represents the number of reacted electrons and N_{ip} is the number of incident photons. This quantity is used to determine how effective a photocatalyst is; however, it can be complex, depending on the experimental reaction conditions. In general, the efficient photocatalyst would have a high quantum yield and give a high rate of gas evolution.

The other important element for an efficient photocatalyst is the spectral range of sunlight absorbed. As UV light has higher photon energy, UV-based photocatalysts will perform

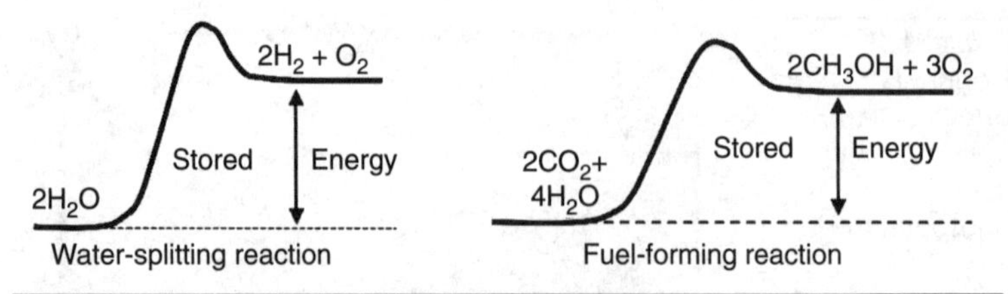

FIGURE 1.3 Photosynthesis by green plants and photocatalytic water splitting as an artificial photosynthesis.

better per photon than visible light-based photocatalysts, but less UV light reaches the earth's surface than visible light from the sun. Thus, even a less efficient photocatalyst that absorbs visible light can be more useful than a more efficient photocatalyst absorbing solely UV light.

The sunlight energy can be converted to chemical energy directly through water splitting, as shown in Figure 1.3. This reaction is similar to photosynthesis by green plants because these are uphill reactions. Therefore, photocatalytic water splitting is regarded as an artificial photosynthesis and is an attractive and challenging theme in chemistry. From the viewpoint of the Gibbs free energy change, photocatalytic water splitting is distinguished from photocatalytic degradation reactions such as photo-oxidation of organic compounds using oxygen molecules, which are generally downhill reactions. This downhill reaction is regarded as a photoinduced reaction and has also been extensively studied using TiO_2 photocatalysts.[22–23] Many research work on photocatalytic water splitting have been published so far.[24–49]

CHAPTER 2

Photocatalytic Reactions, Oxidation, and Reduction

The synthesis of artificial photosynthetic systems to convert H_2O and CO_2 to fuels (for example, CH_3OH) has stimulated new fundamental investigations of the interaction of light, electron flow, and chemical reactions. Photosynthetic processes start with the electron-hole pair formation at the interface of a semiconductor and a solution upon the absorption of sunlight, which leads to the oxidation or reduction of water. Absorption of sunlight depends on the size of the bandgap and positions of the band edges, as well as the overall band structure of semiconducting photocatalysts, which are therefore of crucial importance in photoelectrochemical and photocatalytic applications (Figure 2.1). The energy positions of the band levels can be controlled by the electronegativity of the dopants, solution pH (for example, a flat band potential variation of 59 mV per pH unit), and quantum-confinement effects. Accordingly, the band edges and bandgaps can be tailored to achieve specific electronic, optical, or photocatalytic properties through nanostructure engineering.

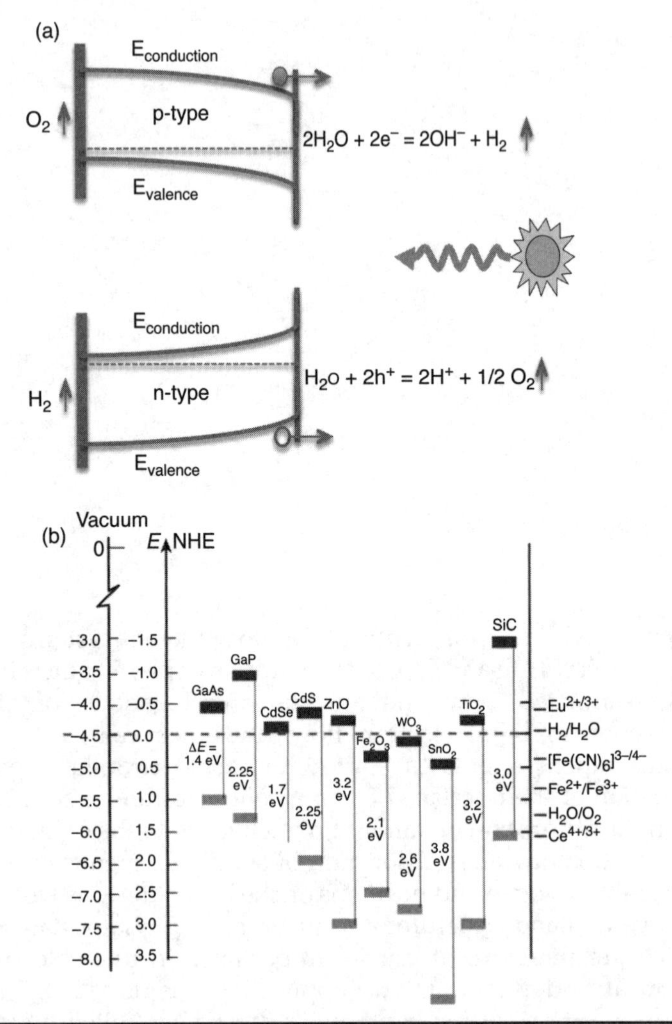

FIGURE 2.1 Important parameters for evaluation of photocatalysts in water splitting.

2.1 Basic Points in Photocatalytic Reactions

Many characterization points should be taken into account when evaluating photocatalytic water splitting.[21]

1. **Stoichiometry.** Water splitting generated H_2 and O_2 gases should be in a stoichiometric ratio of 2:1 when no sacrificial reagent is applied. In many reactions H_2

evolution occurs without O_2 evolution. Sometimes, the amount of H_2 evolution is typically small, so it could be hard to determine whether the reaction is photocatalytic water splitting.

2. **Time course.** The amount of H_2 and O_2 gases generated from water splitting should show an increase with the irradiation time.
3. **Turnover number (TON).** The amount of H_2 and O_2 gases generated should be more than the amount of photocatalyst added. TON is normally defined by the number of reacted molecules to that of an active site [Eq. (2.1)].

$$\text{TON} = \frac{N_{\text{rm}}}{N_{\text{as}}} \tag{2.1}$$

where N_{rm} represents the number of reacted molecules and N_{as} is the number of active sites. In practice the number of reacted electrons to the number of atoms in a photocatalyst [Eq. (2.2)] or on the surface of a photocatalyst [Eq. (2.2)] is often monitored as the TON. The number of reacted electrons can be calculated from the amount of H_2.

$$\text{TON} = \frac{N_{\text{re}}}{N_{\text{atom}}} \tag{2.2}$$

where N_{re} represents the number of reacted electrons and N_{atom} is the number of atoms in a photocatalyst.

$$\text{TON} = \frac{N_{\text{re}}}{N_{\text{surfatom}}} \tag{2.3}$$

where N_{surfatom} is the number of atoms at the surface of a photocatalyst.

4. **Quantum yield.** The photocatalytic activity depends on the experimental conditions, including the light source and reaction cell, so it is difficult to compare the activities between different experiments. In this case, quantum yield becomes an important measure. However, it is hard to determine the real amount of photons absorbed by a photocatalyst in a dispersed

system. So, the obtained quantum yield is an effective quantum yield (EQY) (Eq. 2.4). The EQY will be smaller than the real quantum yield

$$\text{EQY}(\%) = \frac{N_{re}}{N_{ip}} \times 100 \tag{2.4}$$

where N_{ip} is the number of incident photons.

2.2 Semiconductor Catalysts

One of the key challenges is to develop an artificial photosynthetic process with an average efficiency significantly higher than the natural plants or algae (it is worthwhile to point out that the efficiency of the natural photosynthesis process in plants is significantly higher in the early morning when the sun is not strong as compared to noon time; otherwise, plants will not survive in strong sun). An important energy-storage application involves using the solar energy captured in the charge-separated states of a solar-capture-and-conversion system to break and make the chemical bonds, thereby producing solar fuels.[19,50]

The water splitting reaction in Figure 2.2 requires three separate catalytic steps:

1. The positively charged electrode, or anode, removes electrons from hydrogen atoms in water molecules.
2. The hydrogen ions (protons) created break away from their oxygen atoms.
3. The anode catalyst then welds pairs of oxygen atoms together to make O_2.

Meanwhile, the free protons drift through the solution to the negatively charged electrode (cathode), where pairs of protons and electrons combine to make molecular hydrogen (H_2).

This is an example of a general conversion process in a solar fuel cycle that involves evolution of oxygen as one product and formation of a reduced fuel as the other. Unexplored basic science issues are immediately confronted when the problem is posed in the simplest chemistry framework.

An approach to duplicating photosynthesis outside of a photosynthetic membrane is to convert sunlight into spatially separated electron/hole pairs within a photovoltaic cell and

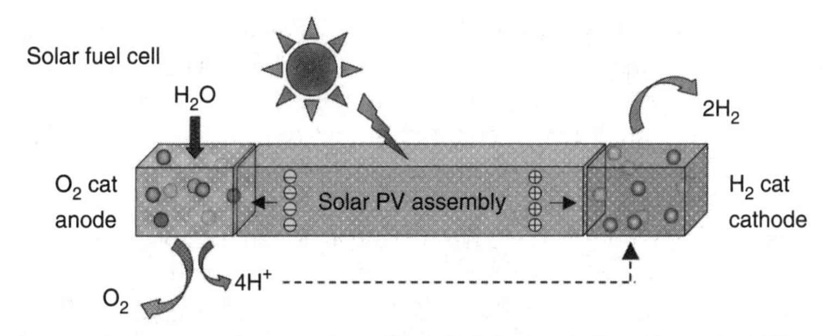

FIGURE 2.2 A fuel cell produces electricity by, for example, passing hydrogen past a catalyst at the anode, producing electrons that flow through a wire and protons that pass to a cathode, where they react with the protons and an oxidant to produce the waste product, such as water. In contrast, the solar fuel cell uses light to reverse the flow of electrons and protons. To effect solar fuel production, the electrons and protons are again coupled to catalysts, but this time they break the water bonds and make H_2 and O_2 (from Ref. 50).

then capture the charges with catalysts that mediate water splitting.[50–51] The four holes are captured by a catalyst at the anode to produce oxygen, and the four electrons are captured by a separate catalyst at the cathode to produce hydrogen. The net result is the storage of solar energy in the chemical bonds of H_2 and O_2. A key determinant of energy storage in artificial photosynthesis is the efficiency of the water-splitting catalysts. Electrocatalysts that are efficient for solar-to-fuels conversion must operate close to the Nernstian potentials (E) for the H_2O/O_2 and H_2O/H_2 half-cell reactions shown in Eq. (2.5) (half-cell potentials given in the convention of reduction potentials).

$$2H_2O \rightarrow O_2 + 4H^+ + 4e \tag{2.5}$$

$$4H^+ + 4e \rightarrow 2H_2 \tag{2.6}$$

$$2H_2O \rightarrow 2H_2 + O_2 \tag{2.7}$$

The voltage in addition to E that is required to attain a given catalytic activity—referred to as overpotential—limits the efficiency of converting light into catalytic current. Of the two reactions, the H_2O/O_2 reaction is considerably more complex.[52] This reaction requires a four-electron oxidation of two water molecules coupled to the removal of four protons to form a relatively weak oxygen-oxygen bond. In addition to controlling this proton-coupled electron transfer (PCET),[53–54] a catalyst must tolerate prolonged exposure to oxidizing conditions. Even at the thermodynamic limit, water oxidation requires an oxidizing power that causes most chemical functional groups to degrade. Accordingly, the generation of oxygen from water presents a substantial challenge toward realizing artificial photosynthesis.[55]

2.3 Biomimetic Catalysts

Finding artificial materials that mimic biomolecules (biomimetic materials) appropriate for artificial solar energy conversion involves answering two questions:

1. Why are biomolecules so successful in separating electrons and holes?
2. How to transfer this capability to simpler and sturdier molecules that can be incorporated into solar cells? The ultimate goal is to find the smallest molecule that still has this capability.

A common example of biomimetic solar energy conversion has been Photosystem II, a widely used prototype for photosynthesis. Figure 2.3 zooms in step by step from a cell membrane in a leaf to the assembly of wheel-shaped protein molecules that collect the electrons and to the active region containing four manganese atoms and one calcium. The mechanism of photosynthesis has been investigated in great

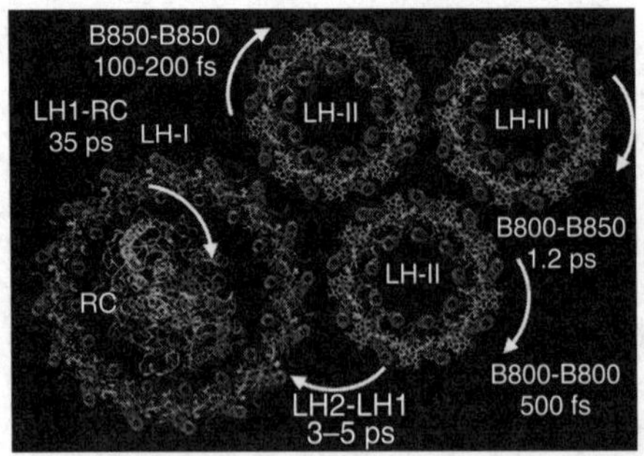

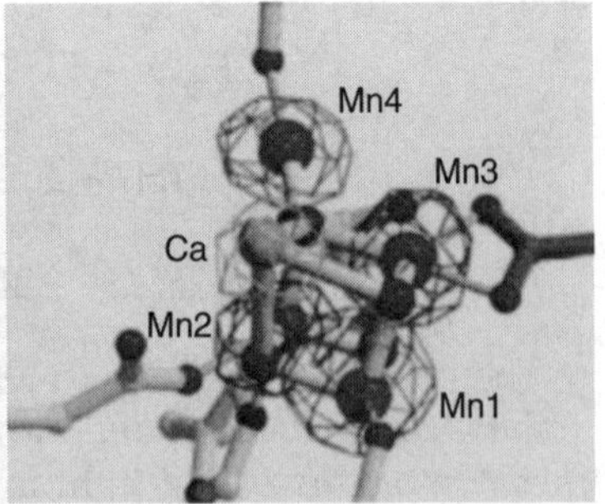

FIGURE 2.3 Photosystem II at various magnifications. The core consists of one calcium atom and four manganese atoms, which change their 3d electron configuration during absorption of light. Spectroscopy of the 2p-to-3d transitions is optimally suited to probe these changes.

detail for model systems such as Photosystem II. The efficiency of photosynthesis in plants is only about 2 percent. However, nature is very efficient in the initial capture of light and the separation of electrons and holes. Biomimetic photosynthesis tries to utilize these tricks of nature.

The electronic structure of the active metal-organic center has been probed by optical and x-ray absorption spectroscopies. Because of the extreme dilution for the possible obtained samples, it has only been possible to observe the K-edge of the 3d transition metals in hard x-ray regions that form the active part of Photosystem II.[56–57] The sharper L-edge absorption that would provide much more chemical information, particularly about the 3d electrons, has not been very successful so far. These are inaccessible from the K-edge, due to dipole selection rules. From the multiplet structure of the L-edge, it is possible to obtain the number of d-electrons, and from the L_2/L_3 branching ratio, the spin state.

2.4 Nanostructured Catalysts

Nanoporous materials provide a path of research toward achieving high densities of small molecules for artificial photosynthetic systems. Nanoporous materials act as efficient electrodes for minimizing the pathway of excitons and/or charge carriers to the collector electrodes and forming a scaffold to hold large numbers of photoactive molecules. In an example of photovoltaic cell, in order to capture a reasonable amount of the incoming light, the layer of dye molecules needs to be made fairly thick, much thicker than the molecules themselves. To address this problem, titanium dioxide nanoparticles are used as a scaffold to hold large numbers of the dye molecules, increasing the number of molecules for any given surface area of cell.

An important example of the use of nanoporous materials is found in the generation of H_2 and O_2 from direct photo-oxidation of water.[7,58–59] To succeed in splitting water using only solar radiation, the valence band maximum of the semiconductor has to be located at a lower energy level than the chemical potential of oxygen evolution (H_2O/O_2), and the conduction band minimum has to be located at a higher energy level than the chemical potential of hydrogen evolution (H_2/H^+). If the valence and conduction band energies do not fulfill this requirement, an external bias has to be applied to induce the photocatalytic process, which in turn substantially reduces the overall efficiency.

CHAPTER 3

Transition Metal Oxides

The inherent complexity associated with the materials engineered for energy technologies remains a challenge both for experimental characterization and theoretical modeling. How can the breakthroughs be achieved? It is increasingly accepted that the design and optimization of novel materials is strongly enhanced by a collaboration between an increased and detailed fundamental understanding of the novel materials and an insight-based establishment of a set of desirable parameters.

Over the last decades scientific and engineering interest in the application of semiconductor photocatalysis has grown exponentially. Semiconductor photocatalysis with a primary focus on transition oxides as a durable photocatalyst has been applied to a variety of problems of environmental interest in addition to water and air purification. It has been shown to be useful for the destruction of microorganisms such as bacteria[60] and viruses,[61] for the inactivation of cancer cells,[62–67] for odor control,[68] for the photosplitting of water to produce hydrogen gas,[69–70] for the fixation of nitrogen[71–73] and for the cleanup of oil spills.[74–76]

Semiconductors (e.g., TiO_2, ZnO, Fe_2O_3, $C2_3O_4$, WO_3, CeO_2, CdS, and ZnS) can act as sensitizers for light-reduced redox processes due to their unique electronic structure, which is characterized by a filled valence band and an empty conduction band.[77] When a photon with an energy of hv matches or exceeds the bandgap energy, E_g, of the semiconductor, an electron, e^-_{cb}, is promoted from the valence band (VB) into the conduction band (CB), leaving a hole, h^+_{vb} behind

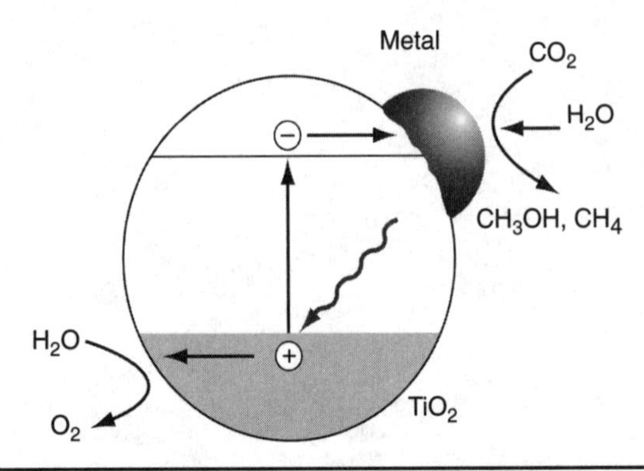

FIGURE 3.1 Primary steps in the photoelectrochemical mechanism. Excited state conduction-band electrons and valence-band holes can recombine and dissipate the input energy as heat, get trapped in metastable surface states, or react with electron donors and electron acceptors adsorbed on the semiconductor surface or within the surrounding electrical double layer of the charged particles.

(see Figure 3.1). The primary steps in the photoelectrochemical mechanism are:

1. Formation of charge carriers by a photon
2. Charge carrier recombination to liberate heat
3. Initiation of an oxidative pathway by a valence-band hole
4. Initiation of a reductive pathway by a conduction-band electron
5. Further thermal (e.g., hydrolysis or reaction with active oxygen species) and photocatalytic reactions to yield mineralization products
6. Trapping of a conduction band electron in a dangling bond to yield, e.g., Ti(III)
7. Trapping of a valence-band hole at a titanol group

In the absence of suitable electron and hole scavengers, the stored energy is dissipated within a few nanoseconds by recombination.[78] If a suitable scavenger or surface defect state is available to trap the electron or hole, recombination is prevented and subsequent redox reactions may occur. The valence-band holes are powerful oxidants (+1.0 to +3.5 V vs. NHE, depending on the semiconductor and pH), while the conduction-band electrons are good reductants (+0.5 to −1.5 V vs. NHE).[79]

Most organic photodegradation reactions utilize the oxidizing power of the holes either directly or indirectly; however, to prevent a buildup of charge one must also provide a reducible species to react with the electrons. In contrast, on bulk semiconductor electrodes only one species, either the hole or electron, is available for reaction due to band bending.[80] However, in very small semiconductor particle suspensions both species are present on the surface. Therefore, careful consideration of both the oxidative and the reductive paths is required.

Hydrogen peroxide is formed on illuminated TiO_2 surfaces in the presence of air via dioxygen reduction by a conduction-band electron in the presence of a suitable electron donor, such as acetate.[81–83] In the presence of organic scavengers, organic peroxides and additional H_2O_2 may be formed through the following generalized sequence.

In most experiments and applications with semiconductor photocatalysts, oxygen is present to act as the primary electron acceptor. As a consequence of the two-electron reduction of oxygen, H_2O_2 is formed via the preceding mechanism. This process is of particular interest since Gerischer and Heller have suggested that electron transfer to oxygen may be the rate-limiting step in semiconductor photocatalysis.[74–85] Hydroxyl radicals are formed on the surface of TiO_2 by the reaction of h^+_{vb}, with adsorbed H_2O, hydroxide, or surface titanol groups (>TiOH). H_2O_2 may also contribute to the degradation of organic and inorganic electron donors by acting as a direct electron acceptor or as a direct source of hydroxyl radicals due to homolytic scission. However, due to the redox potentials of the electron-hole pair, H_2O_2 can theoretically be formed via two different pathways in an aerated aqueous solution as follows:

$$O_2 + 2e^-_{cb} + H^+ \rightarrow H_2O_2 \tag{3.1}$$

$$2H_2O + 2h^+_{vb} \rightarrow H_2O_2 + 2H^+ \tag{3.2}$$

Hoffman et al. and Kormann et al. have shown that the quantum yield for hydrogen peroxide production during the oxidation of a variety of low-molecular-weight compounds has a pronounced Langmuirian (i.e., Langmuir-Hinshelwood) dependence on the O_2 partial pressure. These observations suggest that the primary formation of H_2O_2 occurs via the reduction of adsorbed oxygen by conduction-band electrons. Hoffman et al.[83] have used ^{18}O isotopic labeling experiments to demonstrate that all of the oxygen in photochemically produced hydrogen peroxide (e.g., $H^{18}O^{18}OH$) arises from dioxygen (e.g., $^{18}O_2$) reduction by conduction-band electrons in the case of ZnO photocatalysis of carboxylic acid oxidation. No H_2O_2 is detected in the absence of oxygen.

3.1 Metal Oxide Semiconductors

Several simple oxide and sulfide semiconductors have bandgap energies sufficient for promoting or catalyzing a wide range of chemical reactions of environmental interest. They include TiO_2 ($E_g = 3.2$ eV), WO_3 ($E_g = 2.8$ eV), $SrTiO_3$ ($E_g = 3.2$ eV), α-Fe_2O_3 ($E_g = 3.1$ eV for $O^{2-} \rightarrow Fe^{3+}$ transitions), ZnO (E_g = 3.2 eV), and ZnS (E_g = 3.6 eV). Among these semiconductors TiO_2 has proven to be the most suitable for widespread environmental and energy applications. TiO_2 is biologically and chemically inert, it is stable with respect to photocorrosion and chemical corrosion, and it is inexpensive. The primary criteria for good semiconductor photocatalysts for organic compound degradation are that the redox potential of the H_2O/*OH ($OH^- = {}^*OH + e^-$; E° = −2.8 V) couple lies within the bandgap domain of the material and that they are stable over prolonged periods. The metal sulfide semiconductors are unsuitable based on the stability requirements in that they readily undergo photoanodic corrosion. And the iron oxide polymorphs (α-Fe_2O_3, α-FeOOH, β-FeOOH, γ-FeOOH, and δ-FeOOH) are not suitable semiconductors, even though they are inexpensive and have nominally high bandgap energies, because they readily undergo photocathodic corrosion.[86]

Titanium dioxide in the anatase form appears to be the most photoactive[87–88] and the most practical candidate of the semiconductors for widespread environmental applications, such as water purification, wastewater treatment, hazardous waste control, air purification, and water disinfection. Titanium dioxide is widely used as a white paint pigment, as a sun blocking material, a cosmetic, and a builder in vitamin tablets among many other uses. ZnO appears to be a suitable alternative to TiO_2; however, ZnO is unstable with respect to incongruous dissolution to yield[81,83,89] Zn-$(OH)_2$ on the ZnO particle surfaces, thus leading to catalyst inactivation over time.

In recent years, Degussa P25 TiO_2 has set the standard for photoreactivity in environmental applications, although TO_2 produced by Sachtleben (Germany)[90–91] and Kimera (Finland) show comparable reactivity. Degussa P25 is a nonporous 70:30% anatase-to-rutile mixture with a BET (Brunauer, Emmett and Teller) surface area of 55 ± 15 m^2 g^{-1} and crystallite sizes of 30 nm in 0.1 pm diameter aggregates. Many researchers claim that rutile is a catalytically inactive,[92–94] much less active form of TiO_2,[69,95,96–97] while others find that rutile has selective activity toward certain substrates. Highly annealed (T ≥ 800°C) rutile appears to be photoinactive in the case of 4-chlorophenol oxidation. However, Domenech[98] showed that TiO_2 in the rutile form was a substantially better photocatalyst for the oxidation of CN^- than in the anatase form; on the other hand, he also showed that Degussa P25 was a better catalyst than rutile for the photoreduction of $HCrO_4^-$.[99–100]

Tanaka et al.[101] have shown that photocatalytic degradation of several compounds over different mineral phases and preparation methods of TiO_2 was dependent upon the calcination temperature for some samples and independent for others. They found that the rate of trichloroethene (TCE) photodegradation in water increased with TiO_2 calcination temperatures up to 500°C or in some cases up to 600–700°C and then decreased above those temperatures. They also noted that commercial anatase forms (Degussa and TP-2) were better for trichloroethylene (Cl_2CCClH) degradation than commercially available rutile (Katayama, TP-3 and TM-1) and that specific surface area did not appear to be a determining factor. Tanaka et al. concluded that synthesized anatase that has been calcined was better than P25 and that both of these types were better than 100 percent rutile. However, when hydrogen peroxide was added as an electron acceptor, rutile showed greater photocatalytic activity.

Martin et al.[94] reported an increase in photodegradation rates of 4-chlorophenol as the anatase form of TiO_2 was calcined progressively from 100 to 400°C (i.e., the particles calcined at 400°C yield the highest photodegradation rates) and then a decrease in photodegradation rate was noted for samples calcined above 500°C. For comparison, the apparent quantum efficiency was found to be 0.23 percent for anatase (400°C) and 0.03 percent for rutile (800°C).

3.2 Heterogeneous Photocatalyst Materials

Most metal oxide, sulfide, and nitride photocatalysts consist of metal cations with d^0 and d^{10} configurations. Their conduction bands for the d^0 and d^{10} metal oxide photocatalysts are usually composed of *d* and *sp* orbitals, respectively, while their valence bands consist of O 2*p* orbitals. Valence bands of metal sulfide and nitride photocatalysts are usually composed of S 3*p* and N 2*p* orbitals, respectively. Orbitals of Cu 3*d* in Cu^+, Ag 4*d* in Ag^{4+}, Pb 6*s* in Pb^{2+}, Bi 6*s* in Bi^{3+}, and Sn 5*s* in Sn^{2+} can also form valence bands in some metal oxide and sulfide photocatalysts. Alkali, alkaline earth, and some lanthanide ions usually do not directly contribute to the band formation and simply construct the crystal structure, as A site cations in perovskite compounds.

Some transition metal cations with partially filled *d* orbitals such as Cr^{3+}, Ni^{2+}, and Rh^{3+} form some impurity levels in bandgaps when they are doped or substituted for native metal cations. Although they often work as recombination centers between photogenerated electrons and holes, they sometimes play an important role for visible light response.

Some transition metals and the oxides, such as noble metals (Pt,[102–103] Rh,[103–104] and Au[105–106]), NiO,[107] and RuO_2,[108–109] function as co-catalysts for H_2 evolution. In water splitting, a back reaction to form water from evolved H_2 and O_2 has to be suppressed because of an uphill reaction. In general, Au, NiO, and RuO_2 are suitable co-catalysts on which the back reaction hardly proceeds. A CrRh oxide has recently been found as an excellent co-catalyst for H_2 evolution by oxynitride photocatalysts.[110–111] IrO_2 colloid works as an O_2 evolution co-catalyst.[112–114]

3.3 d^0 Metal Oxide Photocatalyts

TiO_2 is the first reported photocatalyst for water splitting under UV irradiation.[115] TiO_2 can produce hydrogen and/or oxygen from water vapor and pure liquid water, as well as aqueous solutions containing electron donors.[65,104,116–121] It was found that under UV irradiation, colloidal TiO_2, combined with ultrafine Pt and RuO_2 particles, generated H_2 with a high quantum yield of 30 ± 10% and O_2 in stoichiometric proportions from water.[70] The reaction solution had a pH of 1.5, which was adjusted with HCl.[65] The addition of either NaOH or Na_2CO_3 was found to be effective for water splitting using the Pt/TiO_2 photocatalyst.[116,122–123] When TiO_2 was doped with metal ions, the photocatalytic activity for water splitting was effectively enhanced. Chae et al. reported that whereas Ga-doped TiO_2 powder could split pure water stoichiometrically under UV irradiation, pure TiO_2 did not show any activity.[124] The Ni^{2+} doping enhanced the photoactivity of the TiO_2 for hydrogen production from an aqueous methanol solution.[125] Sn/Eu co-doped TiO_2 exhibited a high activity for hydrogen generation with a quantum efficiency of 40 percent with Pd as the co-catalyst under the irradiation from a fluorescent lamp.[126] Zalas et al. studied the effect of lanthanide doping on the photocatalytic activity of TiO_2.[127] The best performance for hydrogen production from an aqueous methanol solution was obtained for the TiO_2 containing 0.5 mol% of Gd oxide as the dopant. The UV-driven photocatalytic activity of TiO_2 was also improved by combining it with a second oxide semiconductor. All of the mixed oxides with heterophase-structures, SnO_2/TiO_2,[128] ZrO_2/TiO_2,[129] Cu_xO/TiO_2,[130–132] Ag_xO/TiO_2,[133–135] and $MTiO_3$/TiO_2 (M = Ca, Sr, Ba),[136] displayed higher rates of photocatalytic hydrogen evolution from aqueous solutions containing electron donors other than TiO_2 alone. With Pt as a co-catalyst, the Ti/B binary oxide stoichiometrically decomposed pure water under UV irradiation.[137–138] When TiO_2 nanoclusters were dispersed in the mesoporous structure of MCM-41 and MCM-48, the formed Ti-MCM-41[139] and Ti-MCM-48[140] showed much higher photocatalytic activity for hydrogen evolution under UV irradiation than bulk TiO_2.

Many white titanates are known to work as efficient photocatalysts for water splitting under UV irradiation. Shibata et al. reported that the layered titanates, $Na_2Ti_3O_7$, $K_2Ti_2O_5$, and $K_2Ti_4O_9$ were active in photocatalytic H_2 evolution from aqueous methanol solutions even without the presence of a Pt co-catalyst.[141] These layered titanates, consisting of titanium oxide layers and interlayers, can be modified using ion-exchange reactions.[141–142] Of the materials studied, the H^+-exchanged $K_2Ti_2O_5$ exhibited a high activity with a quantum yield of up to approximately 10 percent. After being pillared with SiO_2 in the interlayers, $K_2Ti_4O_9$ showed an enhanced photocatalytic activity for H_2 evolution from CH_3OH/H_2O mixtures. This seems to agree with the increase in the surface area.[143] The $Na_2Ti_2O_5$ titanate nanotube material with a nickel complex intercalated into the interlayers also showed a high photocatalytic activity for H_2 evolution from water/methanol solutions under UV irradiation.[144] Kudo et al. found that a range of Cesium compounds, $Cs_2Ti_nO_{2n+1}$ (n = 2, 5, 6), with layered structures showed photocatalytic activities for H_2 and O_2 evolution from aqueous solutions.[145] The $Cs_2Ti_2O_5$ with a five-coordinate structure consisting of TiO_5 units was more active than $Cs_2Ti_5O_{11}$ and $Cs_2Ti_6O_{13}$ with six-coordinate structures consisting of TiO_6 units. The unsaturated coordination state of the five-coordinate structure worked as the active sites of catalytic reactions and contributed to the photoactivity. The photoactivity of $Cs_2Ti_2O_5$ was also greatly enhanced by the H^+-exchange reaction. Inoue and co-workers investigated a series of alkali-metal titanates with a chemical formula of $M_2Ti_nO_{2n+1}$ (M = Na, K, Rb and n = 2, 3, 4, 6) as photochemical water-splitting catalysts.[146–149] The alkaline metal atoms, M, in $M_2Ti_6O_{13}$ showed a great effect on the photocatalytic activity for water splitting. Interestingly, $RuO_2/M_2Ti_6O_{13}$ (M = Na, K, and Rb) with rectangular tunnel structures showed higher photocatalytic activity than $RuO_2/Cs_2Ti_6O_{13}$ with a layered structure. The activity increased in the order of Na > K > Rb > Cs. Kohno et al. reported that in a photocatalytic system of ruthenium oxide-deposited barium titanates ($BaTi_4O_9$, $Ba_2Ti_9O_2$, $Ba_4Ti_{13}O_{30}$, and $Ba_6Ti_{17}O_{40}$), only $RuO_2/BaTi_4O_9$ was active in water decomposition.[150] The pentagonal prism tunnel structure of $RuO_2/BaTi_4O_9$ gave rise to a higher photocatalytic activity than $RuO_2/K_2Ti_4O_9$ with a zigzag layer structure. It is believed that the tunnel structure was responsible for the high dispersion of the RuO_2 particles.[149,151–152]

The method of catalyst preparation also appears to play a role in the final activity. For example, $BaTiO_3$ synthesized with a polymerized complex method exhibited enhanced photocatalytic activity when compared to the materials prepared by the traditional solid-state reaction method.[153] This was due to the larger surface area. Domen and co-workers reported that a NiO-loaded $SrTiO_2$ powder was capable of decomposing pure water as well as water vapor into H_2 and O_2 under UV irradiation.[107,154–158] The activity of the photocatalyst was increased considerably by a pretreatment in H_2 and using a concentrated NaOH solution for the photocatalytic reaction.[155]

The photocatalytic activity of $SrTiO_3$ was also greatly improved by using a modified preparation method[159] or a suitable concentration of metal cations doping (such as La^{3+},[160] Ga^{3+}, and Na^+).[161] Some derivatives, such as $Sr_3Ti_2O_7$ and $Sr_4Ti_3O_{10}$, were also found to be active in water decomposition when loaded with NiO as the co-catalyst.[162–163] Mizoguchi et al. reported that platinized $CaTiO_3$ powder, with bandgap of 3.5 eV estimated from optical absorption edge, also exhibited a high photocatalytic activity under UV irradiation.[164] By doping with Zr^{4+} to form a $CaTi_{1-x}Zr_xO_3$ solid solution, the activity was further increased. Quantum yields of up to 1.91 percent and 13.3 percent for H_2 evolution from pure water and aqueous ethanol solution, respectively, were reported.[165]

Lee and co-workers investigated a series of perovskites whose layers were integrated with the intergrowth with the same elements (La and Ti) but had different layer thicknesses.[166–168] La_2TiO_5, $La_2Ti_2O_7$, and $La_2Ti_3O_9$, which had layered structures made up of slabs of 1, 3, and 4 units, respectively, showed much higher photocatalytic activities under UV irradiation than bulk $LaTiO_3$. Alkaline earth component doping (Ba, Sr, and Ca) was shown to improve the photoactivities of $La_2Ti_2O_7$. In particular, the NiO-modified Ba-doped $La_2Ti_2O_7$ proved extremely active for overall water splitting, with a quantum yield close to 50 percent when alkaline hydroxide was introduced into the reaction system as an external additive.[169] The activity of $La_2Ti_2O_7$ was greatly enhanced by synthesizing the catalyst using the polymerized approach instead of the solid-state reaction method.[170] In contrast, $La_2Ti_2O_7$ and $Ln_2Ti_2O_7$ (Ln = Pr, Nd) with a layered structure was also active for water splitting but exhibited lower activities, with the activity decreasing in the order $La_2Ti_2O_7 \gg PrLaTi_2O_7 > Pr_2Ti_2O_7 > NdLaTi_2O_7 > Nd_2Ti_2O_7$.[171]

Abe's group first reported titanate $R_2Ti_2O_7$ (R = Y, Eu-Lu) with pure cubic-pyrochlore structure as a water-splitting photocatalyst in 2004.[172] NiO_x-loaded $Y_2Ti_2O_7$ demonstrated the most efficient evolution of H_2 and O_2 in a stoichiometric ratio from pure water under UV irradiation. $Y_2Ti_2O_7$ photocatalysts, with better crystallinity and higher activity, were obtained by the addition of excess Y in the polymerized complex synthetic procedure.[173] It was found that the high photocatalytic activity observed for overall water splitting compared to $R_2Ti_2O_7$ (R = Y, Gd) was related to the increased mobility of the electrons and holes caused by the corner-shared octahedral TiO_6 network in these materials.[174] Uno et al. also studied the photocatalytic activities for hydrogen evolution using $Ln_2Ti_2O_7$ (Ln = La, Pr, Nd, Sm, Gd, Dy, Ho, Er, and Yb).[175] However, only $La_2Ti_2O_7$ and $Sm_2Ti_2O_7$ showed detectable photocatalytic hydrogen evolution at a small rate.

Miseki et al. studied the photocatalytic properties of $ALa_4Ti_4O_{15}$ (A = Ca, Sr, and Ba) with a (111) plane-type layered perovskite structure.[176] Of these, NiO_x-modifed $BaLa_4Ti_4O_{15}$ showed the highest activity for water splitting, with a quantum yield of 15 percent at 270 nm. The highly donor-doped (110) layered perovskite $La_4CaTi_5O_{17}$ was found to be an efficient photocatalyst for overall water splitting, with quantum yield as high as 20 percent under UV irradiation.[177] The spontaneously hydrated layered perovskites with a general formula of $A_{2-x}La_2Ti_{3-x}Nb_xO_{10}$ (A = interlayer cations such as K, Rb, Cs; x = 0 1) were found to be efficient photocatalysts for water decomposition under UV irradiation.[178–180] $NiO/Rb_2La_2Ti_3O_{10}$ produced H_2 and O_2 from a RbOH aqueous solution, with quantum yield up to 5 percent. Suitable modification techniques such as co-catalyst loading, using Ni or Au,[181–182] metal-ion doping (Zn^{2+} doping),[183] and alternative synthetic methods[184] all led to the enhanced activity of $K_2La_2Ti_3O_{10}$. The polymerized complex synthesis method was also utilized to prepare the high-purity stoichiometric $KTiNbO_5$ photocatalyst, which demonstrated increased activity after NiO loading.[185] For $KLaTiO_4$, Zr^{4+} doping had a positive effect on the photocatalytic activity in a water splitting reaction, giving rise to a quantum yield as high as 12.5 percent.[186]

Sekine et al. were the first to examine the photocatalytic reactions on the ion-exchangeable layered titanoniobate compounds $CsNbTi_2O_7$ and $CsNbTiO_5$.[187] They found that the H^+-exchanged forms of those compounds work as efficient photocatalysts for H_2 or O_2 evolution from an aqueous solution containing methanol or silver nitrate under UV irradiation. Members of the aurivillius-type layered perovskites $(Bi_2O_2)^{2+}(A_{n-1}B_nO_{3n+1})^{2-}$ (A = Ba, Bi, etc., B = Ti, Nb, etc.), $Bi_4Ti_3O_{12}$, $BaBi_4Ti_4O_{15}$, and Bi_3TiNbO_9 evolved H_2 and O_2 from aqueous methanol and $AgNO_3$ solutions, respectively.[188] He et al. investigated the photocatalytic activity for hydrogen production over another layered perovskite $KBi_3PbTi_5O_{16}$ under UV irradiation.[189]

3.3.1 Ti-, Zr-Based Oxides

Sayama and co-workers were the first to find that the photocatalytic decomposition of pure water proceeded over ZrO_2 powder without any loaded metals as co-catalysts under UV irradiation.[123,190–191] The activity was significantly affected by the pressure of the reaction system, the nature of the additive, and the pH of the solution. Specifically, the addition of Na_2CO_3 or $NaHCO_3$ led to a remarkable increase in the activity and stability of the gas evolution rate. However, the activity decreased when metals such as Pt, Au, Cu, and RuO_2 were loaded onto the surface of the ZrO_2. It was presumed that the large electronic barrier height of the semiconductor-metal junction prevented the migration of electrons from ZrO_2 to the metal. This would lead to the loaded metals possibly blocking the reaction sites on the ZrO_2. Reddy et al. revealed that the photocatalytic activity of ZrO_2 prepared by the precipitation method was highly dependent on the hydrolyzing agent used.[192] The highest photocatalytic activity was obtained for the ZrO_2 with the highest surface area when KOH was used as the hydrolyzing agent. Zou et al. reported that a clean and direct metal-support interface of a NiO-loaded ZrO_2 photocatalyst could be obtained using a plasma method.[193] In photocatalytic reactions, this interface proved more efficient for the charge separation and transfer, which in turn led to the higher photocatalytic activity for water splitting using NiO/ZrO_2 by the plasma treatment than that prepared with the traditional thermal treatment. Compared to the conventional bulk ZrO_2, the photocatalytic water splitting activity was greatly enhanced by the high dispersion of ZrO_2 into the amorphous wall of MCM-41.[194]

$BaZrO_3$ with a cubic perovskite structure produced hydrogen efficiently with a quantum yield up to 3.7 percent from pure water without the assistance of any co-catalysts under UV irradiation.[195] The high photoactivity of $BaZrO_3$ was attributed to the highly negative potential of the photoinduced electrons, the 180° Zr-O-Zr bond angle, and the large dispersion of the conduction band edge composed of Zr 4d orbitals. When the Zr element was partially substituted by Sn, the photocatalytic activity for water splitting was obviously improved.[196] Uno et al. investigated the photocatalytic activities for hydrogen evolution of a series of lanthanide zirconium oxides, Ln_2ZrO_7 (Ln = La, Ce, Nd, and Sm).[197] Under the illumination of a 500-W Xenon lamp, hydrogen gas was clearly evolved in a distilled water suspension of $La_2Zr_2O_7$, $Sm_2Zr_2O_7$, and $Nd_2Zr_2O_7$. On the other hand, $Ce_2Zr_2O_7$ showed no photocatalytic activity because of its lower conduction band level.

After modification with Pt as a co-catalyst, mesoporous zirconium-titanium phosphates demonstrated considerable activity in photocatalytic water decomposition.[198] The H_2 production rate was gradually increased with the addition of Zr, and the maximum H_2 evolution was observed for the $Zr_{0.5}Ti_{0.5}PO_4$ material. Furthermore, the use of sodium carbonate as a pH adjuster was essential and significant for hydrogen generation and also provided stability for the photocatalytic reaction system. A zirconium phosphate/phosphonate compound with quantum yield of 4 percent was developed to produce hydrogen photochemically from water. It was based only on the ultraviolet portion of the spectrum in the presence of a sacrificial reductant Ethylenediaminetetraacetic acid (EDTA).[199]

3.3.2 Nb-, Ta-Based Oxides

Pure Nb_2O_5, with a bandgap of 3.4 eV determined from the threshold of UV-vis absorption, is not active for pure water splitting under UV irradiation.[191] After modification with Pt as a co-catalyst, however, it can efficiently produce H_2 from aqueous solutions containing methanol as an electron donor.[200] Mesoporous Nb_2O_5 synthesized via an evaporation-induced self-assembly method, demonstrated a photocatalytic activity 20 times higher for hydrogen evolution than a bulk Nb_2O_5 without any porosity. Intercalation of In_2O_3 into the mesoporous structure further increased the photoactivity of mesoporous Nb_2O_5 by 2.7 times.[201]

Besides Nb_2O_5, a large number of niobates can produce H_2 and O_2 via water splitting upon UV irradiation. In 1986, Domen and co-workers developed $K_4Nb_6O_{17}$ as the first example of a niobate photocatalyst that showed high and stable activity for H_2 evolution from aqueous methanol solution without any assistance from other materials such as the noble metals.[202–203] This niobate is composed of layers of niobium oxide sheets in which potassium ions are located in two different kinds of interlayers. The potassium ions between the niobium oxide layers can be exchanged with other cations, including transition metal ions. The activities of catalysts exchanged with H^+, Cr^{3+}, and Fe^{3+} ions were higher than the original $K_4Nb_6O_{17}$. In particular the H^+-exchanged $K_4Nb_6O_{17}$ showed the highest activity for H_2 evolution from an aqueous methanol solution with quantum yield was up to approximately 50 percent at 330 nm.[202,204] After modification with NiO,[205–208] Au,[106] Pt,[209–210] and Cs[211] as co-catalysts, $K_4Nb_6O_{17}$ was quite efficient for simultaneous hydrogen and oxygen evolution from pure water. Upon the addition of alkaline hydroxide (KOH, NaOH) into the aqueous solution, the activity of the NiO/$K_4Nb_6O_{17}$ photocatalyst for overall water splitting was enhanced $Rb_4Nb_6O_{17}$, which has the same layered structure, exhibited a high activity for photocatalytic water splitting to form H_2 and O_2 under irradiation. The quantum efficiency at 330 nm was ca. 10 percent in the initial stage of the reaction over the NiO (0.1 wt%)-$Rb_4Nb_6O_{17}$ photocatalyst.[212] When doped with Ta_2O_5, layered compounds of the type $A_4Ta_xNb_{6-x}O_{17}$ (A = K or Rb, $x = 2, 3$ and 4) were able to decompose water stoichiometrically after the pretreatment of H_2 reduction and O_2 reoxidation at high temperatures. Some other alkaline-metal niobates such as $ANbO_3$ (A = Li, Na, K)[213–215] and $Cs_2Nb_4O_{11}$[216] also catalyzed H_2 and/or O_2 evolution from water under UV irradiation, but only after modification with Pt, RuO_2, or NiO. Ikeda et al. found that tungsten-containing alkaline niobates with a defect pyrochlore structure, $ANbWO_6$ (A = Rb, Cs), and loaded with NiO, showed photocatalytic activity for overall water splitting under UV irradiation.[217] The conduction bands of the materials were thought to be composed of the W 5*d* orbital hybridized with the Nb 5*d* orbital.

Photocatalytic water splitting over the alkaline-earth niobates has been studied by various researchers. The related strontium niobates $SrNb_2O_6$,[218] $Sr_2Nb_2O_7$[121,177,219] and $Sr_5Nb_4O_{15}$[176] exhibited efficient photocatalytic activities for hydrogen and oxygen production from pure water under UV irradiation. In particular, $Sr_2Nb_2O_7$ with a highly donor-doped (110) layered perovskite structure gave quantum yields as high as 23 percent.[177] The activity of $Sr_2Nb_2O_7$ was further enhanced to give a quantum yield of 32 percent, by using a hydrothermal synthetic process, which produced a 1D nanostructure with larger BET surface areas.[121] In contrast, the quantum yield of $Ca_2Nb_2O_7$ was 7 percent.[177] $Ba_5Nb_4O_{15}$ with a layered perovskite structure was studied by Kudo and co-workers.[176–177] It gave a 17 percent of quantum yield at 270 nm for water splitting when loaded with NiO co-catalysts. Partial substitution of Nb^{5+} with Zn^{2+} gave the resulting $BaZn_{1/3}Nb_{2/3}O_3$ with a distorted perovskite structure that showed favorable photocatalytic activity under UV irradiation.[220–221]

Domen and co-workers reported a novel Dion-Jacobsen series of ion-exchangeable niobates, $A(M_{n-1}Nb_nO_{3n+1})$ (A = Na, K, Rb, Cs; M = La, Ca, Sr, etc.), with layered perovskite structures that showed unique photocatalytic activities. This was especially true for the H^+-exchanged forms, for H_2 evolution from aqueous alcohol solutions, and for O_2 evolution from an aqueous silver nitrate solution.[222] For example, $KSr_2Nb_3O_{10}$ produced hydrogen at a rate of 0.11 mmol/h/g. After cation exchange with protons, the rate of $HSr_2Nb_3O_{10}$ increased to 43 mmol/h/g. The related layered niobate $K_{2.33}Sr_{0.67}Nb_5O_{14.335}$ was also reported by Wu and co-workers to show much higher photoactivity for hydrogen evolution after a proton exchange reaction.[223] They further reported that other modifications for proton-exchanged $H(M_{n-1}Nb_nO_{3n+1})$ (M = La, Ca, Sr, etc.), such as metal-ion doping (La^{3+},[224] In^{3+}[225], and Mo^{6+}[226] doped into $HCa_2Nb_3O_{10}$ and $H_2LaNb_2O_7$, respectively), efficiently improved the photocatalytic hydrogen evolution from aqueous methanol solutions. When $H_2LaNb_2O_7$ was synthesized using a polymerized complex method it showed higher activity for water splitting than the same material prepared by a solid-state reaction.[227]

Ebina and co-workers synthesized a restacked aggregate of exfoliated nanosheets of $[Ca_2Nb_3O_{10}]^-$ by flocculation with NaOH and KOH aqueous solutions. Under UV irradiation, the restacked aggregates, with a 10-fold enhancement of the surface area, showed higher activities for photocatalytic hydrogen evolution than the $KCa_2Nb_3O_{10}$ starting compound;[228] overall photocatalytic splitting of water was achieved when RuO_x was intercalated between the layers during the exfoliation-restacking route.[229] After exfoliation using tetrabutylammonium hydroxide, the restacked TBA_x $[H_{1-x}Ca_2Nb_3O_{10}]$ sheets loaded with Pt co-catalysts produced hydrogen from pure water with a quantum efficiency of 7.5 percent. However, no oxygen was observed. Transient absorption measurements of the nanosheets revealed charge separation on a nanosecond time scale.[230] SiO_2-pillared

$HCa_2Nb_3O_{10}$, prepared from the layered perovskite $KCa_2Nb_3O_{10}$ via alkylammonium ion-intercalated $HCa_2Nb_3O_{10}$, showed much higher photocatalytic activity of H_2 evolution from aqueous solutions of long chain alcohols.[231–232] This effect was attributed to the expanded interlayer space facilitating the intercalation of such alcohols that could then serve as efficient electron donors. In 2008, a new member of the Dion-Jacobsen perovskites, $H_2Ca_4Nb_6O_{20}$, was reported to be active for H_2 evolution in the presence of methanol as a sacrificial agent under UV irradiation.[233] When Nb was partially substituted by Ta, the resulting $H_2Ca_4Ta_2Nb_4O_{20}$ showed the best photocatalytic activity. The photocatalytic H_2 evolution rate was 8.5 mmol/h/g. Abe and co-workers investigated the effect of crystal structure on water splitting using R_3NbO_7 (R = Y, Yb, Gd, La).[174,234] Only La_3NbO_7, which had an orthorhombic weberite structure to form a NbO_6 octahedral network that increased the mobility of both electrons and holes, was active for the photocatalytic water-splitting reaction. On the other hand, $La_{1/3}NbO_3$, crystallizing in an A-site-deficient, perovskite-type structure instead, catalyzed H_2 evolution with a Pt co-catalyst from CH_3OH/H_2O solution. In addition, O_2 evolution from $AgNO_3$ aqueous solution under UV irradiation was observed.[235]

Kudo et al. reported a new $ZnNb_2O_6$ photocatalyst consisting of d^{10} and d^0 metal ions. [236] Under UV irradiation, the activity of the native $ZnNb_2O_6$ was negligible, whereas NiO-loaded $ZnNb_2O_6$ showed high activity after an H_2-reduction and O_2-oxidation pretreatment. Chen and co-workers prepared a new series of layered perovskite photocatalysts, $ABi_2Nb_2O_9$ (A = Ca, Sr, Ba), by the conventional solid-state reaction method.[237] Under UV irradiation, these photocatalysts showed photocatalytic activity for both H_2 and O_2 evolution from aqueous solutions containing sacrificial reagents (methanol and Ag^+). The activities decreased in the order of $SrBi_2Nb_2O_9 > BaBi_2Nb_2O_9 > CaBi_2Nb_2O_9$. Zou and co-workers found that the compounds M_2BiNbO_7 ($M = In^{3+}, Ga^{3+}$) with pyrochlore structures were sensitive to UV irradiation and had the ability to split water stoichiometrically to produce H_2 and O_2.[238] On the other hand, Bi_2MNbO_7 ($M = Al^{3+}, Ga^{3+}, In^{3+}$) with the same pyrochlore structure only evolved H_2 or O_2 in the presence of CH_3OH or $Ce(SO_4)_2$ as sacrificial agents.[239] When doped with La^{3+}, the photocatalytic hydrogen production over NiO/Bi_2AlNbO_7 was efficiently improved with La^{3+} doping.[240] In comparison, Bi_2MNbO_7 (M = Y^{3+}, Ce^{3+}, Gd^{3+}, Sm^{3+}, Nd^{3+}, Pr^{3+}, and La^{3+}) evolved H_2 at only a small rate from pure water under UV irradiation.[241] The increase of ion radius of M^{3+} in Bi_2MNbO_7 led to the decrease in the photocatalytic activity. The narrower bandgap formed by the smaller ion radius of M^{3+} suggested easier excitation for an electron from the valence band to the conduction band in the oxide semiconductor. $BiNbO_4$ with a triclinic structure produced only a small amount of hydrogen from pure water under UV irradiation. When doped with Ta, the resulting $BiTa_{0.8}Nb_{0.2}O_4$ had an orthorhombic structure and exhibited much higher activity due to the modified band levels and bandgaps.[242]

Ta-based oxides are known to be active photocatalysts for water splitting under UV irradiation. Under UV (4.0 eV) irradiation, Ta_2O_5 alone only produces a very small amount of H_2 and no O_2 from pure water.[243,224] After modification with NiO and RuO_2 as the co-catalysts, it displayed greater activity for the overall water decomposition.[244] The addition of Na_2CO_3 is also effective for improving the photocatalytic activity of Ta_2O_5, as observed for the TiO_2 photocatalytic system.[123] Mesoporous Ta_2O_5 was found by Domen and co-workers to be an active catalyst for photocatalytic water decomposition after NiO loading.[245] Although the walls of the mesoporous Ta_2O_5 were amorphous, the photocatalytic activity was higher than that of crystallized Ta_2O_5. This was because the small wall thickness of mesoporous Ta_2O_5 favored the migration of excited electrons to the surface. When mixed with Ta_2O_5, the mesoporous Ta-Ti-mixed oxides formed showed relatively high activities for photocatalytic hydrogen generation from methanol/water mixtures under the irradiation of $\lambda > 300$ nm. However, a higher concentration of TiO_2 in the mixed oxides led to the destruction of the mesoporous structure.[246] In contrast, the introduction of MgO into the mesoporous structure of Ta_2O_5 improved the thermal stability and also gave rise to stable photocatalytic activity for overall water decomposition over Mg-Ta-mixed oxides.[247] Ni-mixed mesoporous Ta oxide possessed an incomplete mesoporous structure but exhibited higher photocatalytic activity for overall water decomposition than non-modified mesoporous Ta oxide under UV irradiation, while Cu-mixed mesoporous Ta oxide evolved H_2 only at a low rate.[248]

In 1998, Kudo and co-workers reported that alkali tantalates $ATaO_3$ (A = Li, Na, and K) showed high activities for photocatalytic water splitting into H_2 and O_2 under UV irradiation.[244] The excess of alkali in the synthetic process of the solid-state reaction improved the photocatalytic activities of naked tantalates. The order of the activities was $KTaO_3 << NaTaO_3 < LiTaO_3$.[249] Nickel oxide[250–251] and nanosized gold particles were shown to function as efficient co-catalysts for photocatalytic water splitting. Among them, $NiO/NaTaO_3$ was the most photocatalytically active, and produced H_2 and O_2 from pure water with a quantum yield of 28 percent at 270 nm.[252] Compared to the solid-state reaction method, both the hydrothermal method[253–254] and sol-gel method[255] produced $ATaO_3$ (A = K, Na) with good crystallinity as well as high surface areas. Ishihara and coworker found that controlling the charge density in $KTaO_3$ by doping small amounts of acceptors such as tri- or tetravalent cations was effective for improving the photolysis activity.[256–257] In particular, $NiO/KTaO_3$ doped with 8 mol% Zr^{4+} exhibited a higher photocatalytic activity than that of the well-known photocatalyst Pt/TiO_2.

Kudo and coworker investigated the doping of lanthanide (La, Pr, Nd, Sm, Gd, Tb, and Dy)[258] and alkaline-earth metal ions (Ca, Sr, and Ba)[259] into $NaTaO_3$ photocatalysts for efficient water splitting. Lanthanum was the most effective dopant. The apparent quantum yield at 270 nm amounted to 56 percent,[260] which is the highest quantum yield ever reported for catalysts in pure water splitting. The positive effects on the photocatalytic properties were mainly due to the decrease in the particle size and the ordered surface nanostructure. The many characteristic steps created by the doping affected the electron-hole recombination kinetics as revealed by a time-resolved infrared absorption study of the $NaTaO_3$-based photocatalysts.[261–262] The related photocatalysts $NiO/A_2Ta_2O_6$ (A = K, Na) also worked efficiently for overall water splitting under UV irradiation.[263–264] For the alkaline-earth tantalates ATa_2O_6 (A = Ca, Sr, Ba), the order of photocatalytic activities was $SrTa_2O_6 > BaTa_2O_6 > CaTa_2O_6$. This corresponded to the bandgap and the energy of the photogenerated-electron/hole pairs transferring in the crystal (emission energy).[244,260,250] When the NiO co-catalysts were loaded, $NiO/SrTa_2O_6$ showed the highest activity for overall water splitting. The quantum yield was 7 percent at 270 nm. In the related series of strontium tantalates $Sr_mTa_nO_{(m+5n/2)}$, the photocatalytic activities for water splitting into H_2 and O_2 decreased in the following order: $Sr_2Ta_2O_7 > Sr_5Ta_4O_{15} > SrTa_2O_6 > Sr_4Ta_2O_9$.[265] The quantum yield of NiO (0.15 wt %)/$Sr_2Ta_2O_7$ prepared by the polymerized complex method was estimated to be 24 percent at 270 nm.[266] Substitution of Ta by Nb reduced the bandgap of $Sr_2Ta_2O_7$ from 4.5 to 3.9 eV.[266–267] Under UV irradiation, all the $Sr_2(Ta_{1-x}Nb_x)_2O_7$ solid solutions loaded with NiO co-catalysts decomposed water into H_2 and O_2 stoichiometrically. But the photocatalytic activity decreased dramatically even when the amount of Nb was small. $Ca_2Ta_2O_7$ with a pyrochlore structure produced H_2 and O_2 in a stoichiometric ratio under UV irradiation. The activity was higher than NiO-loaded niobate pyrochlore, NiO/$Ca_2Nb_2O_7$.[263] Otsuka et al. claimed that in comparison to $BaTa_2O_6$, $Ba_5Ta_4O_{15}$ prepared under a Ta-rich atmosphere showed a higher photocatalytic activity in the decomposition of H_2O into H_2 and O_2 under UV irradiation.[268] When Ta was partially replaced by Ni or Zn, only H_2 from water was evolved from the resulting $BaM_{1/3}Ta_{2/3}O_3$ (M = Ni, Zn). Methanol was the electron donor and Pt the co-catalyst.[220,269]

Shimizu and coworker developed $A_2ATa_2O_7$ (A = H, K, and Rb; A = Sr and $La_{2/3}$) with a hydrated layered perovskite structure. It proved highly efficient for overall water splitting, even without co-catalysts loading.[270–271] These catalysts showed higher activities than the anhydrous perovskites ($Li_2SrTa_2O_7$, $La_{1/3}TaO_3$, and $KTaO_3$). This was attributed to their hydrated layered structure where the photogenerated electrons and holes can be effectively transferred to the interlayer water. Moreover, as a result of the intercalation of small NiO clusters into the layers of $H_2La_{2/3}Ta_2O_7$ via an ion-exchange reaction, their co-catalyst action remarkably increased the overall activity. This was achieved by shortening the migration distance of the photogenerated charges to the reactive sites. On the other hand, the relatively large NiO particles at the external surface of $H_2SrTa_2O_7$ did not improve the activity. Interestingly, hydration under aqueous conditions changed the crystal structure of $K_2Sr_{15}Ta_3O_{10}$ from orthorhombic to a tetragonal symmetry. This was due to water intercalation into the interlayer space.[272] With RuO_x as the co-catalyst, the hydrous $K_2Sr_{1.5}Ta_3O_{10}$ photocatalyst was active in the pure water splitting process: the quantum yield at 252.5 nm was 2 percent. For the layered perovskites $ACa_2Ta_3O_{10}$ (A = Cs, Rb, K, Na, and Li), interlayer hydration was only observed for A = Na and Li.[273] Hydration of the Li phase doubled the rate of photocatalytic gas evolution when loaded with the co-catalyst NiO, the highest in the $ACa_2Ta_3O_{10}$ series. Analogously, a quantum yield of 8 percent was achieved for the Ba-based (100) layered perovskite $KBa_2Ta_3O_{10}$ after modification with NiO.[177]

Li et al. prepared another hydrated layered perovskite, namely the tantalate $H_{1.81}Sr_{0.81}Bi_{0.19}Ta_2O_7$ from $Bi_2SrTa_2O_9$, using an ion-exchange reaction in hydrochloric acid solution.[274] Under UV irradiation, the $H_{1.81}Sr_{0.81}Bi_{0.19}Ta_2O_7$ photocatalyst showed favorable photocatalytic activity in splitting pure water into H_2 and O^2 even without the assistance of a co-catalyst. Layered lanthanide tantalates and their ion-exchanged phases ($MLnTa_2O_7$, M = Cs, Rb, Na, and H; Ln = La, Pr, Nd, and Sm) were prepared by Machida and coworker to evaluate their photocatalytic activity for water splitting under UV irradiation.[275–278] The photocatalytic activity was sensitive to not only Ln, but also the interlayer cation, M. The highest activity was obtained for M = Rb with a sequence of Ln: Rb > Nd > Sm > La > Pr. The effects of lanthanide ions on the photocatalytic activities of $LnTaO_4$ (Ln = La, Ce, Pr, Nd, and Sm) with monoclinic structures and $K_2LnTa_5O_{15}$ (Ln = La, Pr, Nd, Sm, Gd, Tb, Dy, and Tm) with tungsten bronze structures were also investigated by the Machida's group.[279–280] The photocatalytic activities for water splitting under UV irradiation strongly depended on the particular lanthanide ion. For the $LnTaO_4$ series, $LaTaO_4$ showed the highest rate for the stoichiometric evolution of H_2 and O_2 from pure water, while among $K_2LnTa_5O_{15}$ series, $K_2PrTa_5O_{15}$ and $K_2SmTa_5O_{15}$ showed relatively high activities for water splitting ($h\upsilon > 3.9$-4.1 eV).

The photocatalytic activity of R_3TaO_7 (R = Y, Yb, Gd, La) for water splitting was studied by Arakawa and coworker.[174,234] They examined the effect of R^{3+} ionic radius on the crystal structure.[174,234] The crystal structures of R_3TaO_7 changed with increasing ionic radius of the R^{3+} ion from a fluorite-type cubic structure to a pyrochlore-type cubic structure and finally to a weberite-type orthorhombic structure. In the case of the La_3TaO_7, the photocatalytic activity was greatly increased by the phase transition from cubic to orthorhombic. $K_3Ta_3Si_2O_6$ and $K_3Ta_3B_2O_{12}$, which have similar crystal structures consisting of pillars formed by a corner sharing of three linear TaO^6 chains, were active for water splitting without any co-catalyst.[281–282] The TaO_6 pillars in $K_3Ta_3Si_2O_{13}$ and $K_3Ta_3B_2O_{12}$ are linked by Si_2O_7 ditetrahedral units and BO_3 triangle units, respectively. Thus, the bond angle of O-Ta-O in $K_3Ta_3B_2O_{12}$ (171.5°) is slightly smaller than that in $K_3Ta_3Si_2O_{13}$ (173.1°). The distortion due to the smaller bond angle of $K_3Ta_3B_2O_{12}$ than that found in $K_3Ta_3Si_2O_{13}$ resulted in a wider bandgap and higher photocatalytic activity. NiO co-catalyst loading increased the activity of $K_3Ta_3Si_2O_{13}$ drastically, but proved ineffective for $K_3Ta_3B_2O_{12}$. In a manner similar to niobates, the tantalates $ATaWO_6$ (A = Rb, Cs) crystallized into a defect pyrochlore structure with a conduction band composed of the W5d orbital hybridized with the Ta 4*d* orbitals. Under UV irradiation, they evolved H_2 and O_2 in a stoichiometric ratio from an aqueous AOH (A = Rb, Cs) solution loaded with NiO as the co-catalyst. Even though the pyrochlore-like Bi_2MTaO_7 (M = Y and La) showed a strong optical absorption in the visible region, as photocatalyts, these oxides could only produce H_2 and O_2 from pure water under UV irradiation.[283]

Chen and coworker investigated the photocatalytic water splitting of La_2AlTaO_7 with a view to studying the effect of aluminum on the electronic structure.[284] It was found that instead of the Ta 5d state it was the Al 3*s*3*p* states that acted as the lower conduction band. Under UV irradiation, La_2AlTaO_7 showed photocatalytic activity in splitting pure water into H_2 and O_2 even in the absence of a co-catalyst. In comparison, using $ABi_2Ta_2O_9$ (A = Ca, Sr, Ba), also developed by Chen's group, H_2 or O_2 evolved only from the aqueous solution containing either methanol or $AgNO_3$ as the sacrificial reagent. Simultaneous production of H_2 and O_2 from pure water was not observed.[285] The photocatalytic activities decreased in the order of $SrBi_2Ta_2O_9 > CaBi_2Ta_2O_9 > BaBi_2Ta_2O_9$. A large range of transition-metal tantalates has been investigated by different researchers with water decomposition as the aim. Under UV irradiation, $NiTa_2O_6$ produced both H_2 and O_2 from pure water without a co-catalyst. On the other hand, with $MnTa_2O_6$, $CoTa_2O_6$, $CrTaO_4$, $PbTa_2O_6$, $FeTaO_4$, and $BiTaO_4$, only traces of H_2 evolved.[242,244] When NiO was loaded as the co-catalyst, $AgTaO_3$ and $ZnTa_2O_6$ were active in water splitting, producing both H_2 and O_2.[244,286] In contrast, in an aqueous methanol solution, $Sn_2Ta_2O_7$ with Pt as the co-catalyst only produced H_2, while $SnTa_2O_6$ was totally inactive.[287]

3.3.3 W-, Mo-Based Oxides

The number of heterogeneous photocatalysts based on either tungstates or molybdates for H_2 or O_2 evolution is quite small. Some were found to be active for water splitting only under UV irradiation even though they showed optical absorption in the visible region. Inoue and coworker found that $PbWO_4$ incorporating a WO_4 tetrahedron showed high and stable photocatalytic activity for the overall splitting of water. A stoichiometric quantity of H_2 and O_2 was produced under UV irradiation when RuO_2 was loaded onto the metal oxide.[288–289] The photocatalytic performance was attributed to large dispersions in both the valence and conduction bands. This generated very mobile photoexcited holes and electrons. In contrast, a small dispersion in the conduction band was observed for the photocatalytically inactive $CaWO_4$, which has a similar crystal structure. $PbMoO_4$, which is related, catalyzed hydrogen evolution from aqueous methanol solution. It also was capable of oxygen evolution from aqueous silver nitrate solution under UV irradiation—oxygen evolution activity comparable to that observed on TiO_2.[290]

Kudo and coworker have extensively investigated the photocatalytic activities of tungstates and molybdates. Under UV irradiation, $Na_2W_4O_{13}$[145] and $Bi_2W_2O_9$,[188] with layered structures, were active for photocatalytic hydrogen (Pt as co-catalyst) and oxygen evolution in the presence of suitable sacrificial reagents. However, Bi_2MoO_6 with a similar structure evolved only oxygen from $AgNO_3$ aqueous solution at a low rate.[188] They also found that some scheelite-type molybdates and tungstates functioned as photocatalysts for both H_2 and O_2 evolution in the presence of sacrificial reagents.[291] The band-gaps of scheelite compounds narrowed when they were composed of Ag^+ and Bi_{3+} ions. $(NaBi)_{0.5}$ MoO_4 (BG: 3.1 eV), $(AgBi)_{0.5}$ WO_4 (BG: 3.2 eV), and $(AgBi)_{0.5}$ MoO_4 (BG: 3.0 eV) all showed photocatalytic activity for O_2 evolution from an aqueous solution containing an electron acceptor. On the other hand, $(NaBi)_{0.5}$ WO_4 (BG: 3.5 eV) produced H_2 from an aqueous solution containing an electron donor. In comparison, $(AgLn)_{0.5}$ MoO_4 (Ln = La, Ce, Eu, Yb) barely produced H_2 or O_2 from aqueous solutions under UV irradiation.

Nguyen et al. developed a novel silicotungstic acid (SWA)-SiO_2 photocatalyst by impregnating SWA on a silica support.[292] Under UV irradiation, an approximately stoichiometric production ratio of H_2 and O_2 was observed on these SWA-SiO_2 photocatalyts. The roles of the photoactive sites in SiO_2 as a donor source for hydrogen formation and that of SWA as an inhibitor for the recombination of photoexcited electrons and holes were appreciable and responsible for the superior photo catalytic performance of the SWA-SiO_2 system.

3.3.4 Other d^0 Metal Oxides

Some other miscellaneous d^0 metal oxides that can catalyze water splitting to H_2 and/or O_2 under UV irradiation are described below. Wang et al. observed that a new crystal structure for nanostructured VO_2, with a body centered-cubic structure (bcc) and a large optical bandgap of $\approx$2.7 eV, surprisingly showed excellent photocatalytic activity in hydrogen production from a solution of water and ethanol under UV irradiation.[293] The bcc VO_2 phase exhibited a high quantum efficiency of 38.7% when synthesized as nanorods. Luan and cowokers prepared Bi_2GaVO_7 and Bi_2YVO_8 with tetragonal structures by solid-state reactions. These two compounds initiated both H_2 and O_2 evolution from pure water only under UV irradiation. This is in spite of the fact both of them showed strong optical absorption in the visible region ($\lambda > 420$ nm).[294–295]

3.4 d^{10} Metal Oxide Photocatalyts

Various typical metal oxides with d^{10} (In^{3+}, Ga^{3+}, Ge^{4+}, Sn^{4+}, Sb^{5+}) configurations have all shown to be effective photochemical water-splitting catalysts under UV irradiation. Of these, Ni-loaded Ga_2O_3 was one of the promising photocatalysts for overall water splitting.[296] Its photocatalytic activity could be effectively improved by the addition of Ca, Cr, Zn, Sr, Ba, and Ta ions.[297] In particular, Zn ion doping remarkably improved the photocatalytic activity—an apparent quantum yield for Ni/Zn-Ga_2O_3 of around 20 percent. By combining with Lu_2O_3, the resulting Zn-doped Lu_2O_3/Ga_2O_3 became a novel composite photocatalyst for stoichiometric water splitting under UV irradiation. When the system was loaded with NiO as the co-catalyst, the quantum yield at 320 nm was estimated to be 6.81 percent.[298] For solid solutions consisting of Ga_2O_3 and In_2O_3, $Ga_{1.14}In_{0.86}O_3$ showed the highest photocatalytic activity for H_2 evolution from aqueous methanol solutions and for O_2

evolution from aqueous silver nitrate solutions.[299] In comparison, with the solid solutions of Y_2O_3 and In_2O_3, $Y_{1.3}In_{0.7}O_3$ showed the highest photocatalytic activity for the overall water splitting when combined with RuO_2 as a promoter.[300]

Inoue and coworker investigated the photocatalytic properties for water decomposition of alkali metal, alkaline earth metal, and lanthanum indates with an octahedrally coordinated In^{3+} d^{10} configuration ion.[301–305] The photocatalytic activity for water decomposition under UV irradiation was considerably large for RuO_2-dipsersed $CaIn_2O_4$, $SrIn_2O_4$, and $Sr_{0.93}Ba_{0.07}In_2O_4$ but very poor for RuO_2-dispersed $AInO_2$ (A = Li, Na) and $LnInO_3$ (Ln = La, Nd). The geometric structures of the InO_6 octahedral units for these indates were compared. It was shown that the photocatalytically active indates possessed distorted InO_6 octahedra with dipole moments. The internal fields that arose due to the dipole moment promoted the charge separation in the initial process of photoexcitation. In addition, the broad *sp* conduction bands with large dispersions permitted the photoexcited electrons to move to the dispersed promoter RuO_2 particles. A group of *p*-block metal oxides was reported to have stable activity in decomposing water to H_2 and O_2 under UV irradiation when combined with RuO_2 or Pt as a co-catalyst.[304–311] They consist of metal ions with d^{10} configurations and have distorted octahedral and/or tetrahedral structures. For example, the distorted SbO_6 octahedra in $M_2Sb_2O_7$ (M = Ca, Sr),[306] $CaSb_2O_6$,[306] and $NaSbO_3$[306], the distorted GeO_4 tetrahedra in Zn_2GeO_4[308], and the distorted InO_6 octahedra and GeO_4 tetrahedra in $LiInGeO_4$[309] were dominantly responsible for photocatalytic activity for water decomposition. Some other metal oxides with d^{10} configuration, such as $ZnGaO_4$,[307] Sr_2SnO_4,[304] and $SrSnO_3$[310–311] were also reported to show photocatalytic activity for water splitting.

3.5 f^0 Metal Oxide Photocatalysts

The f-block metal oxides, usually combined with other metal oxides as photocatalysts. Pure CeO_2 powder was reported to show a consistent activity toward O_2 production in aqueous solutions containing Fe^{3+} and Ce^{4+} as electron acceptors.[312] Sr^{2+}-doped CeO_2 was an active photocatalyst for overall water splitting when RuO_2 was loaded as a promoter.[313] Ce(III) oxides supported zeolites that showed higher photocatalytic activity for pure water splitting.[314] Non-stoichiometric H_2 and O_2 evolution was observed. Photoirradiation of Ce^{3+} species generated electrons ($Ce^{3+} + h\upsilon \rightarrow Ce^{4+} + e^-$) that were captured effectively by water molecules for the production of hydrogen. Yuan et al. reported that $BaCeO_3$ produced H_2 and O_2 from aqueous solutions containing CH_3OH and $AgNO_3$ sacrificial reagents, respectively. It also showed some activity under UV irradiation for overall water splitting with the aid of RuO_2 loading.[315]

CHAPTER 4

Crystal Structure and Electronic Structure

4.1 Crystal Structure

Transition metal (TM) oxides, especially 3d series, are important materials in many technological applications. For instance, in the chemical industry, the TM oxides are the functional components in the catalysts used in many processes to convert hydrocarbons to other chemicals. The TM oxides are also used as electrode materials in electrochemical processes. In the electronics industry, TM oxides are used to make conductors in films. The recently discovered high-temperature superconductors (HTS) and colossal magnetoresistance (CMR) materials are a series of multicomponent transitional metal oxides.

Among these applications, perhaps the use of transition metal oxides as catalysts is the most technologically advanced and economically important, yet most challenging. It is also a research area in which much progress has been made in recent years in terms of understanding the fundamental processes that occur, primarily because advances in instrumentation and experimental techniques have made it possible to study the chemistry of the interface between the transition metal oxide and the gaseous and fluid phase in greater details than ever before. In particular, developments in surface science techniques have provided detailed pictures about the surface structures, chemical compositions, and electronic properties of the reaction surfaces.

Oxides commonly studied as catalytic materials belong to the structural classes of corundum, rocksalt, wurtzite, spinel, perovskite, rutile, and layer structure. These are the structures often reported for the oxides prepared by common methods under mild conditions. In some cases, other structures exist. The positions of the ions may not be at the ideal positions of the highest symmetry. For example, distortions are found for FeO, NiO, MnO, and CoO from the cubic lattice, and VO_2, NbO_2, MoO_2, and WO_2 from the perfect rutile structure.[316–317]

The rocksalt structure comprises a three-dimensional array of alternating cations and anions (Figure 4.1a). Each ion is at the center of an octahedron in which vertices are ions of the opposite type. The structure can be viewed as being made up of corner-sharing octahedra (Figure 4.1b). The wurtzite structure is made up of a three-dimensional net of corner-sharing tetrahedra (Figures 4.1c and 4.1d), Where, each ion is in the center of a tetrahedron in which the opposite ions are at the vertices.

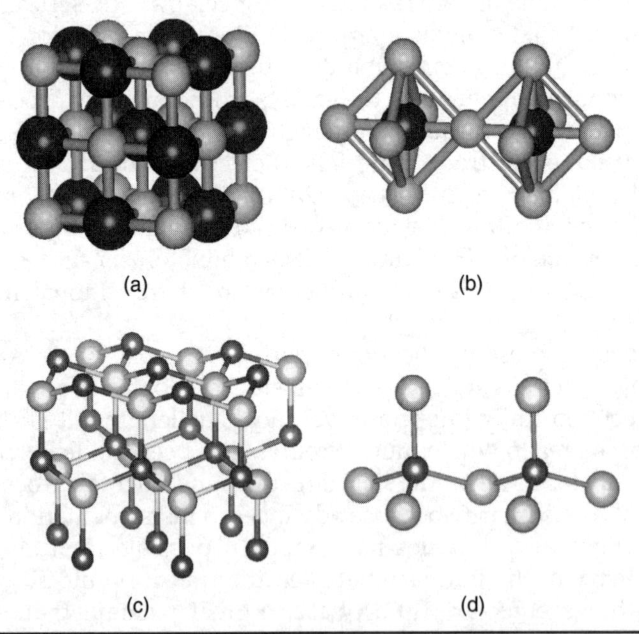

FIGURE 4.1 (a) A rocksalt structure; (b) Two corner-sharing octahedra, one centered around ion A. Octahedra centered around ions A and B are edge-sharing; (c) A wurtzite structure; (d) Two corner-sharing tetrahedra.

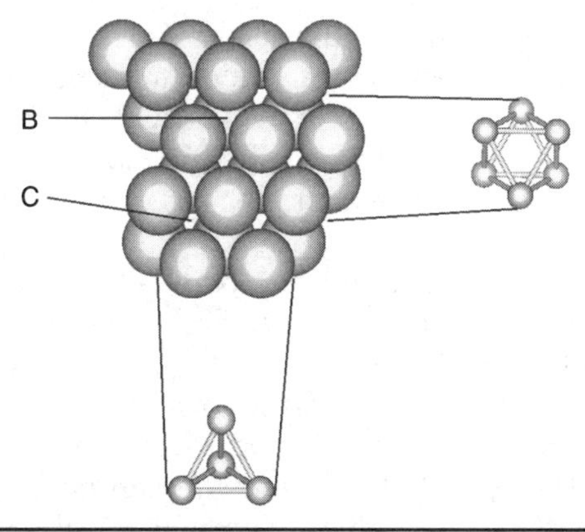

FIGURE 4.2 Close-packed layers of oxide ions. If ions in the third layer are above B, it is a hexagonal close-packed structure. If the ions are above C, it is a cubic close-packed structure. Ions that define octahedral and tetrahedral holes are also shown.

The corundum, the rutile, and the spinel structures are made up of layers of close-packed oxygen ions. The oxygen ions are modeled by hard spheres, and metal ion in a close-packed layer is in contact with six others (see Figure 4.2). When an ion in this layer is in contact with the maximum possible number of ions in the other layer, this ion will be sitting above a triangular hole of the other layer (point B or C), and it will be in contact with three ions in the other layer. For a case of two close-packed layers stacked as shown in Figure 4.2, when a third layer is added on top of these two, the ions take the positions vertically on top of the ions in the first layer such as above point B, or the positions above point C. For the case of point B, the spatial positions of the layers follow the sequence ababab . . . The resulting structure is called hexagonal close packing (HCP), and it forms the basis for the corundum and the rutile structure. In the case of point C, the spatial sequence of the layers is abcabcabc . . . The resulting structure is cubic close packing (CCP), and it forms the basis for the spinel structure.

Between adjacent layers of oxygen ions in both HCP and CCP, the interstices (holes) are bound by either four or six oxygen ions (Figure 4.2). The structures are commonly called tetrahedral and octahedral holes, respectively. There are the same number of octahedral holes as that of oxygen ions, and half as many tetrahedral holes as octahedral holes. In an ideal rutile structure, half of the octahedral holes are filled with cations, while the tetrahedral holes are empty. Thus, the compound has a formula MO_2 (e.g., TiO_2). In the corundum structure, two-thirds of the octahedral holes are filled. The tetrahedral holes are empty, and the compound has a formula M_2O_3(e.g., α-Fe_2O_3).

An ideal spinel structure has one-half of the tetrahedral holes and one-half of the octahedral holes filled, and the formula is M_3O_4 (e.g., Fe_3O_4). It is clear that for charge neutrality, the cations must have two different oxidation states. The most common oxidation states are 2+ and 3+, and the formula can be written as $M^{II}M^{III}O_4$. Both M^{II} and M^{III} can occupy either the tetrahedral or octahedral holes. A normal spinel is one in which all M^{II} ions are in the tetrahedral holes, and all M^{III} ions in the octahedral holes. $ZnFe_2O_4$, an inverse spinel, has all M^{II} ions in the octahedral holes. The M^{III} ions are distributed equally for the octahedral and the tetrahedral holes. Fe_3O_4 and $MgFe_2O_4$, the mixed spinels, have intermediate distributions.

γ-Fe_2O_3 is a spinel. The compound has fewer cations than needed to complete an ideal spinel structure. Fe_3 $(Fe_{5\phi})O_{12}$ is used to represent the fact that for every 12 oxygen ions, there are three Fe^{III} ions in the tetrahedral holes and five Fe^{III} ions in the octahedral holes. Compared to the ideal spinel structure, the occupancy of the tetrahedral holes is the same, but the occupancy of the octahedral holes is one-sixth less. This sixth position is denoted by ϕ to represent a cation vacancy. So, there is one cation vacancy for every three spinel units. When this vacancy is ordered, the repeating unit becomes a trispinel.

The spinel and the corundum structures can be viewed in a different way. The same structure can be constructed using the octahedral units as building blocks instead of constructing the solid with close-packed layers of oxygen ions. Oxygen ions are at the corners of the octahedral units, and cations are at the centers. The corundum structure is made up of a three-dimensional network of such octahedra in which some octahedra share corners, edges, or faces, whereas the spinel structure is made up of octahedra that share corners and edges.

For the rutile structure, the sheets of close-packed oxygen ions are rather distorted. The cations are at the center of octahedra of oxygen ions, as shown in Figure 4.3(a). Along the c-direction (the vertical direction), the octahedra are linked by

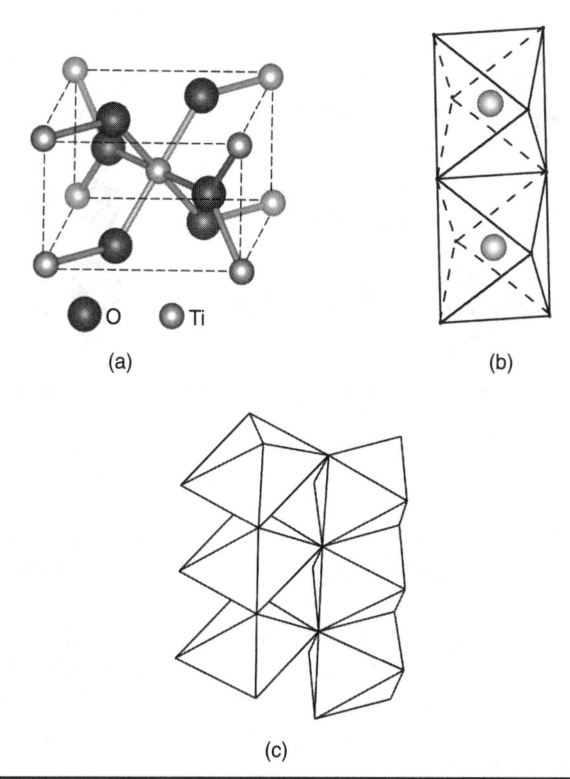

Figure 4.3 (a) An unit cell of rutile TiO_2; (b) Two edge-sharing octahedra of TiO_6 units from adjacent unit cells; (c) A network of octahedra that makes up TiO_2.

sharing edges (Figure 4.3(b)) to form a chain. Adjacent chains are connected by sharing vertices (Figure 4.3(c)). For TiO_2, the octahedra are distorted such that four metal-oxygen distances are of one value, and the other two are of a different value. For some others, like the dioxides of V, Nb, Mo, and W, the metal-metal distances along the octahedra chain are not regular.

Compounds of the perovskite structure are usually written as $M_IM_{II}O_3$. An ideal perovskite structure is made up of a cubic net of corner-sharing octahedra (Figure 4.4). The smaller and more highly charged cation, M_{II} sits in the center of an octahedron, and the larger cation, and less charged cation, M_I sits in the center of the cavity defined by a cube of eight octahedra. Thus, this latter cation is coordinated with 12 oxygen ions. Typically, the M_{II} ions are the transition metal

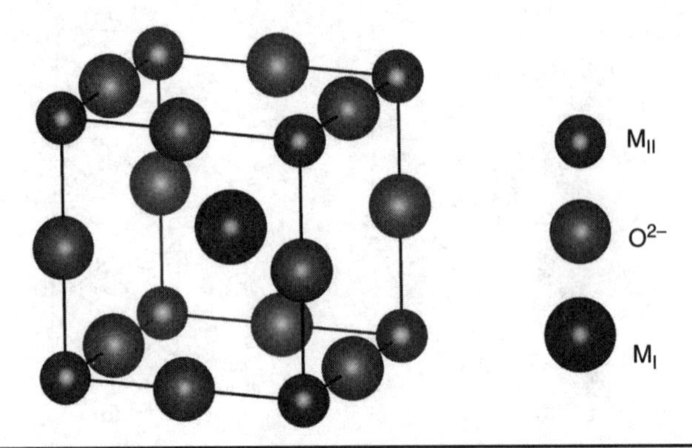

FIGURE 4.4 An unit cell of perovskite $M_IM_{II}O_3$.

ions, and the M_I ions are the alkali, alkali earth, or lanthanide ions. Some examples of perovskites are $KTaO_3$, $SrTiO_3$, and $LaCoO_3$.

There are many other structures, such as scheelite, pyrochlore, and wolframite etc, and the variety increases with the number of components in the compound. The readers are referred to the text by Wells[316] for more detailed discussions of the crystal structures.

4.2 TiO_2

Since its commercial production in the early twentieth century, titanium dioxide (TiO_2) has been widely used as a pigment[317] and contents in sunscreens,[318–319] paints,[320] ointments, toothpaste,[321] and so on. In 1972, Fujishima and Honda demonstrated the reaction of photocatalytic splitting of water on a TiO_2 electrode using ultraviolet (UV) light.[7,22,322] Since then, enormous efforts have been devoted to the research of solar energy conversion into the chemical fuels in higher efficiency. The research on TiO_2 material has also led to many promising applications in areas ranging from photovoltaics and photocatalysis to photo-/electrochromics and sensors.[323–327] These applications can be roughly divided into energy and environmental categories, many of which depend not only on the properties of the TiO_2 material itself, but also on the engineered ligands of the TiO_2 material hosted (e.g., with

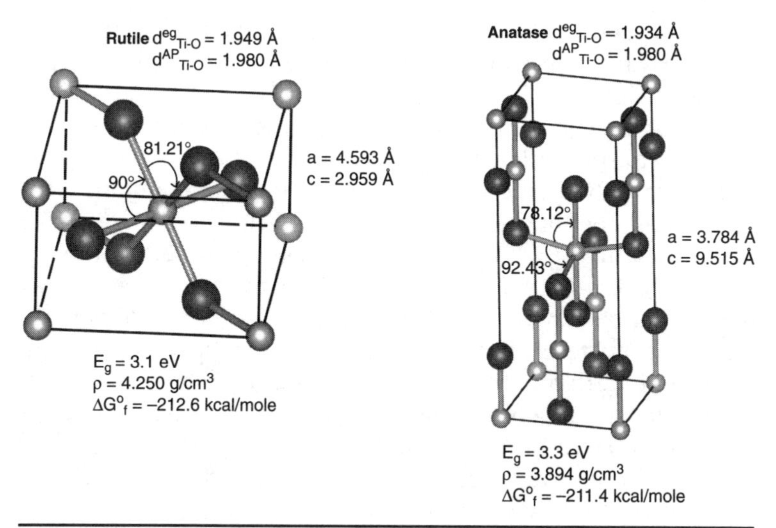

Figure 4.5 Crystal structure of rutile and anatase of TiO_2 (from Ref. 325).

inorganic and organic dye molecules) and on the interactions between the TiO_2 materials and the environment.

Titanium dioxide has promising properties for the energy and environmental applications.[7,323,328] Most available anatase TiO_2 crystals are dominated by the thermodynamically stable 101 facets (more than 94 percent, according to the Wulff construction[329]), rather than the much more reactive 001 facets.[329–337]

Figure 4.5 shows the unit cell structures of the rutile and anatase TiO_2.[325] Both structures can be described in terms of chains of TiO_6 octahedra, where each Ti^{4+} ion is surrounded by an octahedron of six O^{2-} ions. These two crystal structures differ in the distortion of each octahedron and by the assembly pattern of the octahedra chains. In rutile, the octahedron shows a small orthorhombic distortion; in anatase, the octahedron is significantly distorted so that its symmetry is lower than orthorhombic. The Ti-Ti distances in anatase are larger, whereas the Ti-O distances are shorter than those in rutile. In the rutile structure, each octahedron is in contact with 10 neighboring octahedrons (two edge-sharing oxygen pairs and eight corner-sharing oxygen atoms), while, in the anatase structure, each octahedron is in contact with eight neighbors (four sharing an edge and four sharing a corner).

These differences in lattice structures cause different mass densities and electronic band structures between the two forms of TiO_2.

Rutile is a stable phase at the higher temperatures, but anatase and brookite are common in fine-grained (nanoscale) natural and synthetic samples. On heating concomitant with coarsening, the following transformations are all seen: anatase to brookite to rutile, brookite to anatase to rutile, anatase to rutile, and brookite to rutile. These transformation sequences imply very closely balanced energetics as a function of particle size. The surface enthalpies of the three polymorphs are sufficiently different that crossover in thermodynamic stability can occur under conditions that preclude coarsening, with anatase and/or brookite stable at small particle size.[338,339] However, abnormal behaviors and inconsistent results are occasionally observed.

Hwu et al. found the crystal structure of TiO_2 nanoparticles depended largely on the preparation method.[340] For small TiO_2 nanoparticles (<50 nm), anatase seemed more stable and transformed to rutile at >973 K. Banfield et al. found that the prepared TiO_2 nanoparticles had anatase and/or brookite structures, which transformed to rutile after reaching a certain particle size.[341] Once rutile was formed, it grew much faster than anatase. They found that rutile became more stable than anatase for particle size >14 nm. Ye et al. observed a slow brookite to anatase phase transition below 1053 K along with grain growth, rapid brookite to anatase and anatase to rutile transformations between 1053 K and 1123 K, and rapid grain growth of rutile above 1123 K as the dominant phase.[342] It was concluded that brookite could not transform directly to rutile but had to transform to anatase first. However, direct transformation of brookite nanocrystals to rutile was observed above 973 K by Kominami et al.[343]

In a later study, Zhang and Banfield found that the transformation sequence and thermodynamic phase stability depended on the initial particle sizes of anatase and brookite in their study on the phase transformation behavior of nanocrystalline aggregates during their growth for isothermal and isochronal reactions.[339] They concluded that, for equally sized nanoparticles, anatase was thermodynamically stable for sizes <11 nm, brookite was stable for sizes between 11 and 35 nm, and rutile was stable for sizes >35 nm.

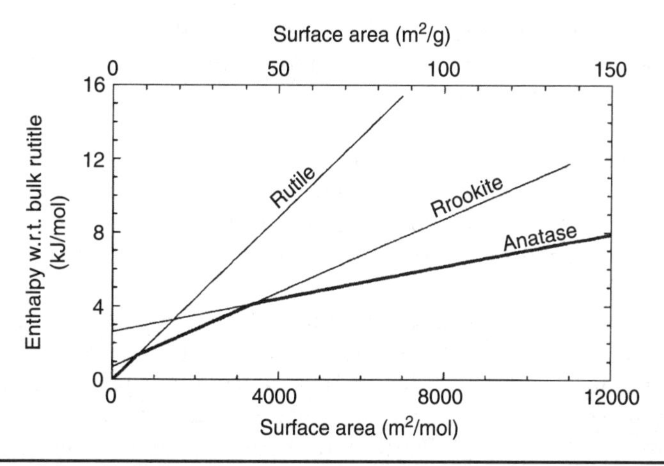

FIGURE 4.6 Enthalpy of nanocrystalline TiO_2 (from Ref. 344).

Ranade et al. investigated the energetics of the TiO_2 polymorphs (rutile, anatase, and brookite) by high-temperature oxide melt drop solution calorimetry, and they found the energetic stability crossed over for the three phases as shown in Figure 4.6.[344] The thick solid line represents the phases of lowest enthalpy as a function of surface area. Rutile was energetically stable for surface area <592 m^2/mol (7 m^2/g or >200 nm), brookite was energetically stable from 592 to 3174 m^2/mol (7–40 m^2/g or 200–40 nm), and anatase was energetically stable for greater surface areas or smaller sizes (<40 nm). The anatase and rutile energetics cross at 1452 m^2/mol (18 m^2/g or 66 nm). Assuming spherical particles, the calculated average diameters of rutile and brookite for a 7-m^2/g surface area were 201 and 206 nm, and those of brookite and anatase for a 40-m^2/g surface area are 36 and 39 nm. These differences in particle size at the same surface area existed because of the differences in density. If the phase transformation took place without further coarsening, the particle size should be smaller after the transformation. Phase stability in a thermodynamic sense is governed by the Gibbs free energy ($\Delta G = \Delta H - T\Delta S$) rather than the enthalpy. Rutile and anatase have the same entropy. Thus, the $T\Delta S$ will not significantly perturb the sequence of stability seen from the enthalpies. For nanocrystalline TiO_2, if the initially formed brookite had surface area >40 m^2/g, it was metastable with respect to both anatase and rutile, and the sequence brookite to anatase to rutile during coarsening was energetically downhill. If anatase formed initially, it could coarsen and transform first to brookite (at 40 m^2/g) and then to rutile. The energetic driving force for the latter reaction (brookite to rutile) was very small, explaining the natural persistence of coarse brookite. In contrast, the absence of coarse-grained anatase was consistent with the much larger driving force for its transformation to rutile.[344]

Li et al. found that only anatase to rutile phase transformation occurred in the temperature range of 973 to 1073 K.[345] Both anatase and rutile particle sizes increased with the increase of temperature, but the growth rate was different, as shown in Figure 4.7. Rutile had a much higher growth rate than anatase. The growth rate of anatase leveled off at 800°C. Rutile particles, after nucleation, grew rapidly, whereas anatase particle size remained practically unchanged. With the decrease of initial particle size, the onset transition temperature was decreased. An increased lattice compression of anatase happened with increasing temperature. Larger distortions existed in samples with smaller particle size. The values for the activation energies were 180 kJ/mol for 12 nm TiO_2 nanoparticles. The decreased thermal stability in finer

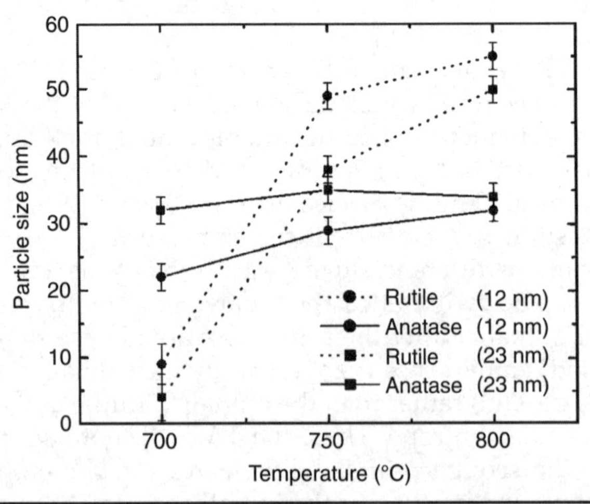

FIGURE 4.7 Changes in particle sizes of anatase and rutile phases of TiO_2 as a function of the annealing temperatures (from Ref. 345).

nanoparticles was primarily attributed to the reduced activation energy as the size-related surface enthalpy and stress energy increased.

4.3 Fe Oxides

α-Fe_2O_3 (hematite) is an antiferromagnetic insulator with some peculiar properties. The ground state configuration of Fe in α-Fe_2O_3 is a high spin d^5 state, for example, ^6S. From a certain transition temperature (T_t = 260 K) to the Néel temperature of 956 K, it shows weak ferromagnetism and its conductivity is activated at high temperatures with a thermal gap of about 2 eV. α-Fe_2O_3 belongs to the transition-metal sesquioxides αM_2O_3 (M = Ti, V, Cr, or Fe) with the α-Al_2O_3 corundum structure. This makes it possible to study and compare electronic properties in a series of transition metal oxides that all possess the same structure but with differing physical properties.

Alumina (α-Al_2O_3) with the corundum structure (Figure 4.8) has a rhombohedral cell with a = 5.136 Å and α = 55.28°, space group R-3c, which contains two formula units. There are four

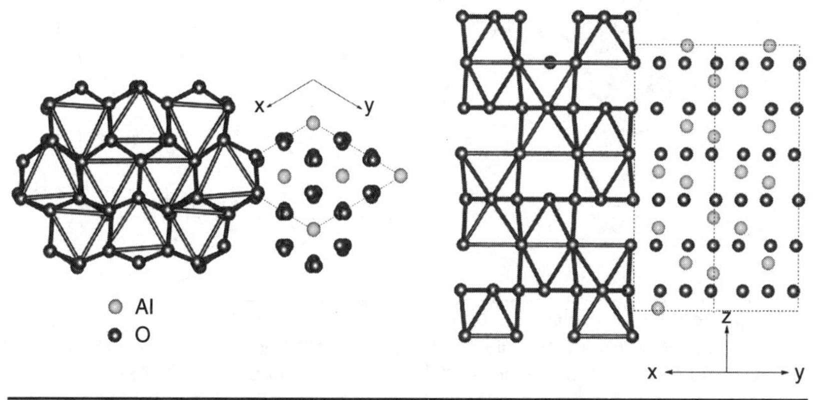

Figure 4.8 The rhombohedral crystal structure of α-Al_2O_3 (space group R3-c). Pairs of face-sharing octahedra extend along the threefold axis. Octahedra of adjacent pairs share corners and edges. Such packages are indicated by braces. On the right side such a layer is projected along the [110] direction.

Al ions at the Wyckoff positions c (0.352, 0.352, 0.352) and six O ions at positions e (0.556, 0.944, 1/4).[346] Each Al ion has three nearest-neighbor oxygen ions at a distance of about 1.85 Å and three next-nearest oxygen ions at a distance of about 1.98 Å.

4.4 Electronic Structure of 3d Transition Metal Oxides

4.4.1 TiO_2

The density of states (DOS) of TiO_2 comprises of Ti e_g, Ti t_{2g} (d_{yz}, d_{zx}, and d_{xy}), O p_σ (in the Ti_3O cluster plane), and O p_π (out of the Ti_3O cluster plane), as shown in Figure 4.9A.[347] The upper valence bands can be decomposed into three main regions:

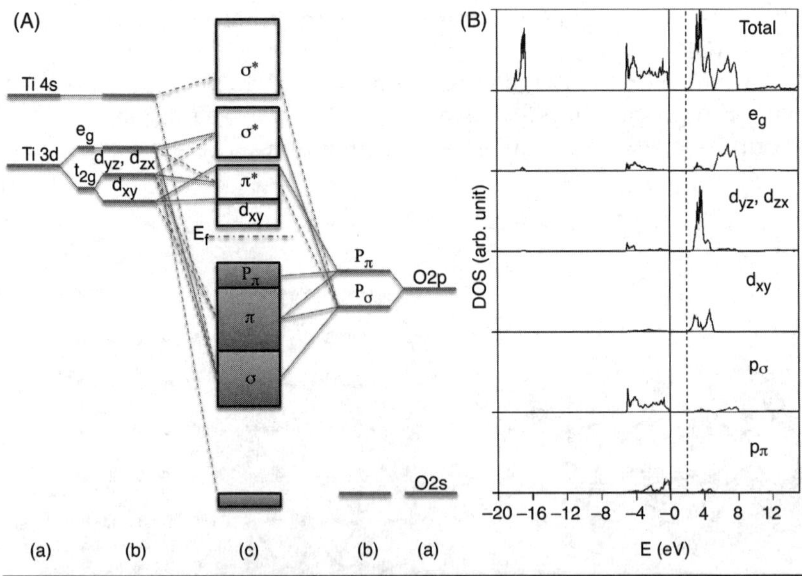

Figure 4.9 (A) Molecular-orbital bonding structure for anatase TiO_2: (a) atomic levels, (b) crystal-field split levels, and (c) final interaction states. The thin-solid and dashed lines represent large and small contributions, respectively; (B) Total and projected densities of states (DOS) of the anatase TiO_2 structure. The DOS is decomposed into Ti e_g, Ti t_{2g} (d_{yz}, d_{zx}, and d_{xy}), O p_σ (in the Ti_3O cluster plane), and O p_π) (out of the Ti_3O cluster plane) components (from Ref. 347).

the σ bonding in the lower energy region mainly due to O p_σ bonding; the π bonding in the middle energy region; and O p_π states in the higher energy region due to O p_π nonbonding states at the top of the valence bands where the hybridization with d states is almost negligible. The contribution of the π bonding is much weaker than that of the σ bonding. The conduction bands are decomposed into Ti e_g (>5 eV) and t_{2g} bands (<5 eV). The d_{xy} states are dominantly located at the bottom of the conduction bands. The rest of the t_{2g} bands are antibonding with p states. The main peak of the t_{2g} bands is identified to be mostly d_{yz} and d_{zx} states.

In the molecular-orbital bonding diagram in Figure 4.9B, a noticeable feature can be found in the nonbonding states near the bandgap: the nonbonding O p_π orbital at the top of the valence bands and the nonbonding d_{xy} states at the bottom of the conduction bands. A similar feature can be seen in rutile; however, it is less significant than in anatase.[348] In rutile, each octahedron shares corners with eight neighbors and shares edges with two other neighbors, forming a linear chain. In anatase, each octahedron shares corners with four neighbors and shares edges with four other neighbors, forming a zigzag chain. Thus, anatase is less dense than rutile. Also, anatase has a large metal-metal distance of 5.35 Å. As a consequence, the Ti d_{xy} orbitals at the bottom of the conduction band are quite isolated, while the t_{2g} orbitals at the bottom of the conduction band in rutile provide the metal-metal interaction with a smaller distance of 2.96 Å.

The electronic structure of TiO_2 has been studied with various experimental techniques, that is, with x-ray photoelectron and x-ray absorption and emission spectroscopies.[340,349–356] Figure 4.10 shows a schematic energy level diagram of the lowest unoccupied MOs of a $[TiO_6]^{8-}$ cluster with O_h, D_{2h} (rutile), and D_{2d} (anatase) symmetry and the Ti L-edge and O K-edge XAS spectra for rutile and anatase.[357] The anatase structure is a tetragonally distorted octahedral structure in which every titanium cation is surrounded by six oxygen ions in an elongated octahedral geometry (D_{2d}). The further splitting of the 3d levels of Ti^{4+} due to the asymmetric crystals is shown for rutile and anatase structures. The fine electronic structure of TiO_2 can be directly probed by Ti L-edge and the O K-edge XAS spectra.[357]

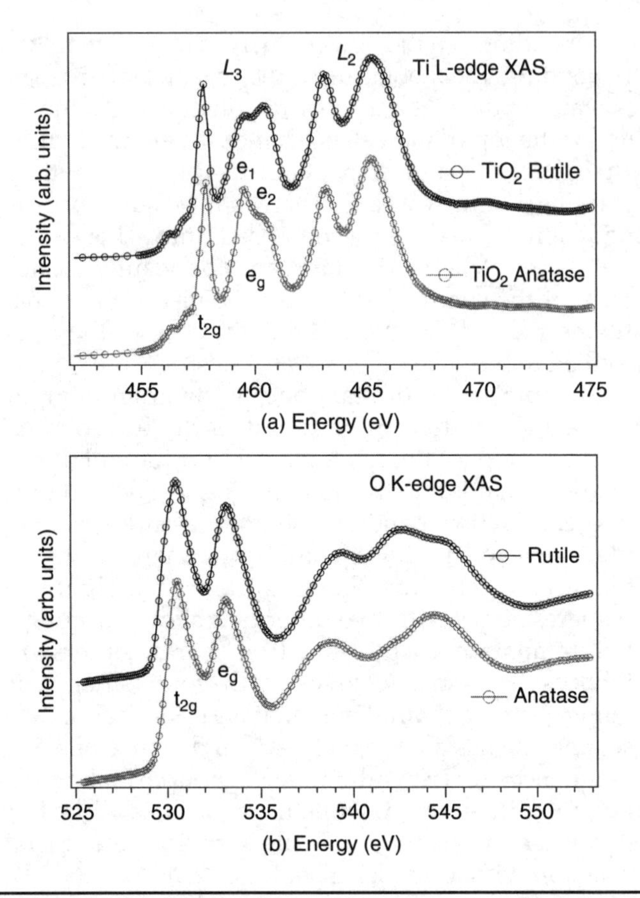

FIGURE 4.10 Ti L-edge and O K-edge XAS spectra for rutile (a) and anatase (b) TiO_2 (from Ref. 357).

4.4.2 ZnO

Figure 4.11 shows the calculated the density of states (DOS) based on a full-potential linearized augmented plane wave method[358] with the LDA exchange-correlation of Perdew and Wang.[359] The Zn $3d^{10}4s^2$ and O $2s^22p^4$ orbitals are treated as valence states. Today's GW methods cannot produce a self-consistent potential, and, therefore, the localization of the d states is corrected by means of the LDA + U^{SIC},[360–362] that is, the LDA with an on-site Coulomb potential for the cation d states. This self-interaction correction (SIC) has shown to result in accurate electronic and optical properties of various

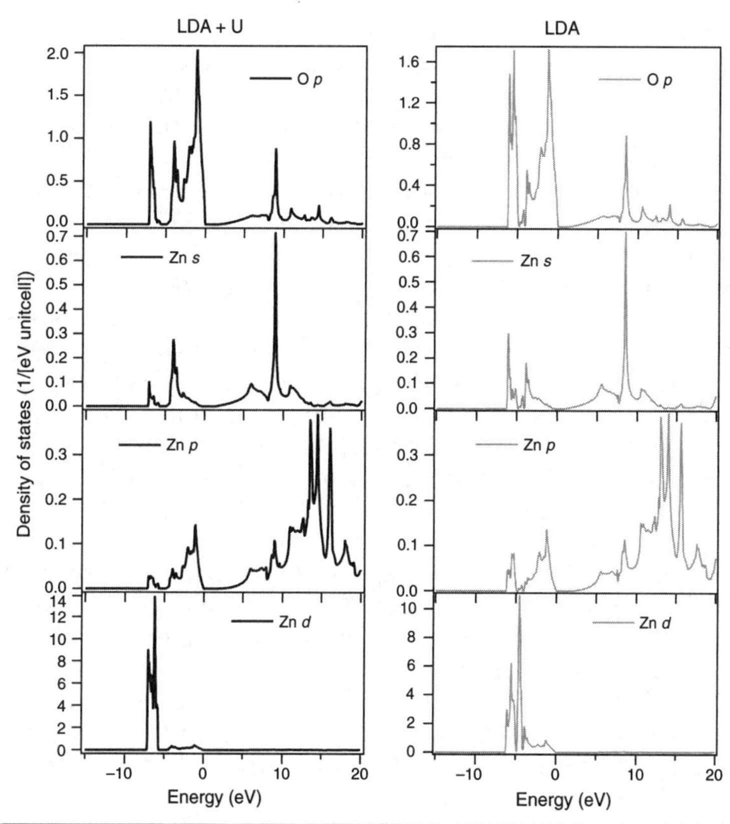

FIGURE 4.11 The partial density of states of ZnO from LDA and LDA + U^{sic} calculations including the O 2p, Zn 4p, 4s, and 3d bands.

s-p hybridized semiconductors.[363] When considering U-d as a fitting parameter, choosing U_d(Zn) = 6.0 eV lowers the G-point Zn 3d states by 0.7 to 1.1 eV, in accordance with the GW results.[364] The different energy corrections of the G-point 3d states (i.e., 0.7–1.1 eV) are due to the change in the hybridization with O p states when the on-site correction potential is applied. In a simple picture, only the Zn 3d t_2-like states (and not the e-like states) couple with the O p states. The LDA + U^{SIC} gives a more narrow band width of the Zn 3d states compared to the LDA method. The calculated O K-edge XES and XAS spectra can be obtained from the DOS and the matrix elements of the dipole-allowed transitions. The modified tetrahedron k-space integration method and a

Lorentzian broadening of 0.6 eV has been used.[358] The relatively large k mesh of 76 k points in the irreducible Brillouin zone is required to ensure convergence of the low-energy dipole-allowed transitions, since both the conduction-band minimum and the valence-band maximum are very nonparabolic near the G point. However, this is less critical in the LDA + U^{SIC} calculation than in the LDA calculation because the LDA + U^{SIC} method yields normally improved band curvatures with more parabolic energy dispersion.

The partial DOS from LDA as well as LDA^{SIC} calculations is displayed in Figure 4.11. Both O 2p and Zn 3d wave functions are strongly localized and the O 2p states are very energetically close to the Zn 3d states, which implies a significantly strong interaction between O 2p and Zn 3d in ZnO. The energy gap of ZnO is primarily a result of the O 2p – Zn 4s interaction, which shifts the O p-like states downward and the Zn s-like states upward. The O 2p states are thus shifted even closer to the Zn 3d levels. Therefore, strong p – d hybridization is expected in ZnO. In addition, the DOS from calculated LDA shows nearly no energy gap between Zn 4s and Zn 3d state, whereas the DOS from calculated LDA + U^{SIC} exhibits the clear energy gap at between –5 and –6 eV in the valence band. Thus, by comparing the calculations and experiments, the locations of Zn 3d energy states can be reproduced fairly well by employing the LDA + U^{SIC} calculation.

CHAPTER 5

Optical Properties and Light Absorption

5.1 TiO_2

The main mechanism of light absorption in pure semiconductors is direct interband electron transitions. This absorption is especially small in indirect semiconductors, for example, TiO_2, where the direct electron transitions between the band centers are prohibited by the crystal symmetry. Braginsky and Shklover have shown the enhancement of light absorption in small TiO_2 crystallites due to indirect electron transitions with momentum nonconservation at the interface.[367] This effect increases at a rough interface when the share of the interface atoms is larger. The indirect transitions are allowed due to a large dipole matrix element and a large density of states for the electron in the valence band. A rapid increase in the absorption takes place at low photon energies ($h\upsilon < E_g + W_c$, where W_c is the width of the conduction band). Electron transitions to any point in the conduction band become possible when $h\upsilon = E_g + W_c$. Further enhancement of the absorption occurs due to an increase of the electron density of states in the valence band only. Such an interface absorption becomes the main mechanism of light absorption for the crystallites that are smaller than 20 nm.[367]

Sato et al. showed through calculation and measurement that the bandgap of TiO_2 nanosheets was larger than the bandgap of bulk TiO_2, due to lower dimensionality, that is, a 3D to 2D transition, as shown in Figure 5.1.[365,368] From the measurement, it was found that the lower edge of the conduction band for the TiO_2 nanosheet was approximately 0.1 V higher, while the upper edge of the

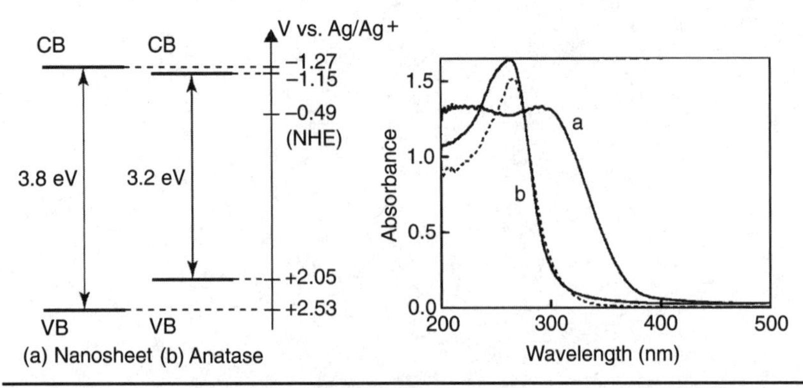

FIGURE 5.1 Schematic illustration of electronic band structure: (a) TiO_2 nanosheets; (b) anatase (from Ref. 365); UV-visible spectra of (a) TiO_2 sheets and (b) a film of nanosheets on a SiO_2 glass substrate (from Ref. 366).

valence band was 0.5 V lower than that of anatase TiO_2. The absorption of the TiO_2 nanosheet colloid blue shifted (>1.4 eV) relative to that of bulk TiO_2 crystals (3.0–3.2 eV), due to a size quantization effect, accompanied with a strong photoluminescence of well-developed fine structures extending into the visible light regime.[366,369] The bandgap energy shift, ΔE_g, by exciton confinement in anisotropic two-dimensional crystallites is formulated as follows:

$$\Delta E_g = \frac{h^2}{8\mu_{xz}}\left(\frac{1}{L_x^2}+\frac{1}{L_z^2}\right)+\frac{h^2}{8\mu_y L_y^2} \tag{5.1}$$

where h is Planks constant, μ_{xz} and μ_y are the reduced effective masses of the excitons, and L_x, L_y, and L_z are the crystallite dimensions in the parallel and perpendicular directions with respect to the sheet, respectively. Since the first term can be ignored, the blue shift is predominantly governed by the sheet thickness. The onset of a 270-nm peak in the photoluminescence of TiO_2 nanosheets was assigned to resonant luminescence. The series of peaks extending into a longer wavelength region were attributed to interband levels generated by the intrinsic Ti site vacancies. The sharp peaks were also attributed to the subnanometer thickness and its uniformity.[366]

Bavykin et al. studied the optical absorption and photoluminescence of colloidal TiO_2 nanotubes with internal diameter in the range of 2.5 to 5 nm, and they found that, in spite of the different diameters, all the TiO_2 nanotubes had similar optical properties.[370] They attributed this to the complete smearing of all one-dimensional effects due to the large effective mass of charge carriers in TiO_2, which resulted in an apparent 2D behavior of TiO_2 nanotubes.

Within the effective mass model, the energy spectrum of 2-D TiO_2 nanosheets can be described by Eq. (5.2), where the plus and minus signs correspond to the conduction and valence bands, respectively; E_G is the energy gap; p is Plancks constant; and m_e and m_h are the effective masses of the electrons and holes, respectively.

$$E_{2D}^{\pm} = \pm\frac{E_G}{2} \pm \frac{h^2k^2}{2m_{e,h}} \tag{5.2}$$

Figure 5.2 shows the absorbance spectra of as-prepared samples of neat TiO_2 and of TiO_2 doped with different amounts of Fe(III). From Figure 5.2 By use of the procedure given by

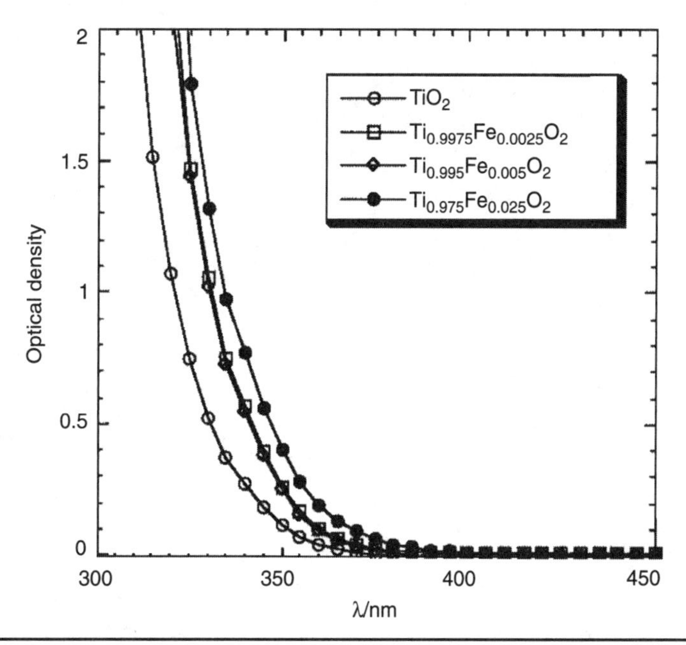

FIGURE 5.2 Absorbance spectra of TiO_2 and iron-doped TiO_2 nanoparticles in colloidal aqueous solution (as prepared at pH = 3, virtually no light scattering) (from Ref. 744).

Kormann et al.[371] the bandgap energy, E_g, of the colloidal particles was obtained as 3.32, 3.25, 3.22, and 3.07 eV for 0, 0.25, 0.5, and 2.5 atom% iron doping, respectively. Clearly, E_g decreases with increasing iron content. Compared with bulk anatase TiO_2 (E_g= 3.23 eV)[372] the bandgap of the undoped TiO_2 particles (3.32 eV) is larger by ca. 0.1 eV. This corresponds to a mean spherical particle diameter of ca 2.7 nm when the Brus equation[373] is applied.

CHAPTER 6

Impurity, Dopants, and Defects

The optical response of a material is largely determined by its underlying electronic structure. The electronic properties of the material are closely related to its chemical composition, atomic arrangement, and physical dimension for nanometer-sized materials. The electronic properties of TiO_2 can be altered by doping. Specifically, the metal (titanium) or the nonmetal (oxygen) component can be replaced in order to alter the material's optical properties. In many applications, it is desirable to maintain the integrity of the crystal structure of the photocatalytic host material and to produce favorable changes in the electronic structure. It appears easier to substitute the Ti^{4+} cation in TiO_2 with other transition metals, and it is more difficult to replace the O^{2-} anion with other anions due to differences in charge states and ionic radii. The small size of the nanoparticle is beneficial for the modification of the chemical composition of TiO_2 due to the higher tolerance of the structural distortion induced by the inherent lattice strain in nanomaterials.[374, 375]

With a view to developing photocatalytic applications using visible light irradiation, beginning with the photocatalytic splitting of water into H_2 and O_2 using TiO_2 under UV irradiation,[7] prophase studies were also carried out on some narrow bandgap semiconductors such as CdS[376,377] and WO_3.[378–380] However, the fact that serious photocorrosion of CdS was observed in the photocatalytic reaction[64,381,382] and the relatively positive conduction band of WO_3 proved dissatisfactory for hydrogen production[379] created the major impediments for the efficient performance of these two photocatalysts for use in visible-light-driven water splitting. Some studies were carried on to improve the photocatalytic stability of CdS,[64,383–386] while others used WO_3 as the photoelectrode in a photoelectrochemical cell to satisfy the energy requirement for water splitting with an applied potential.[387–393] In order to overcome these obstacles, many efforts have been made to develop new visible-light-driven photocatalysts with high water splitting activities. There have been a number of common approaches adopted to make photocatalysts visible-light-active to split water to produce hydrogen and/or oxygen gases:

1. Doping with metal and/or nonmetal ions to control the bandgap and band levels
2. Developing solid solutions to control the band structure
3. Sensitizing dye molecules or quantum dots to make UV-light-active photocatalysts harvest visible light
4. Developing novel single-phase Vis-active photocatalysts through bandgap engineering

6.1 Single-Element Doped TiO_2

6.1.1 Metal Doped TiO_2

One of the most effective ways to develop visible-light-driven photocatalysts is to create impurity levels in the forbidden band through metal ion doping. This makes the wide bandgap photocatalysts active in the visible light region, and this approach has been known for a long time. There have been numerous reports on the modification of wide bandgap photocatalysts using metal ion doping to make them visible-light-active. These include doped TiO_2,[394–397] doped $SrTiO_3$,[397,398] and doped $La_2Ti_2O_7$,[399] among others. Figure 6.1 illustrates the principle of visible-light-driven photocatalysts by metal ion doping to create active photocatalysts with wide bandgaps. A donor level above the original valence band or an acceptor level below the original conduction band, is created to make the photocatalysts respond to visible light.

In 1982, Borgarello et al. found that Cr^{5+} doped TiO_2 could produce hydrogen and oxygen via sustained water cleavage under visible light (400 to 550 nm) irradiation.[402] Since then, many different metal ions have been doped into TiO_2 to improve the visible light absorption and photocatalytic

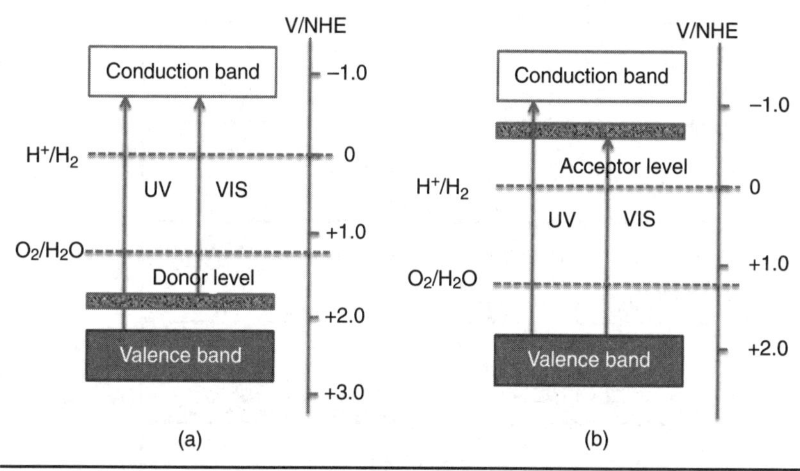

FIGURE 6.1 Donor level (a) and acceptor level (b) formed by metal ion doping.

activities. These include ions such as V, Cr, Mn, Fe, Ni, Mo, and Sn, et al.[394–408] The electronic structure of TiO_2 compounds doped with the 3*d* transition metals (V, Cr, Mn, Fe, Co, and Ni) was analyzed by Umebayashi et al. using *ab initio* band structure calculations (Figure 6.2A).[409] It was found that the 3*d* metal doping created either an occupied level in the bandgap or VB due to the t_{2g} state of the dopant. The charge-transfer transition between this t_{2g} level and the CB (or VB) of TiO_2 contributed to the photoexcitation under visible light. Using DV-Xα calculations, Nishikawa et al. showed that it is possible to shift the absorption edge of titania to the sunlight region in the case of V-, Cr-, Mn-, Fe-, Co-, Ni-, or Rh-doping and revealed the relationship between the ionic radius and the change of the bandgap (Figure 6.2B).[410] And it was found that the cations Ni^{3+} and V^{5+} reduced the bandgap most effectively. Cao et al. reported that Sn^{4+}-doped TiO_2 nanoparticles prepared by the CVD method displayed a higher photocatalytic activity than pure TiO_2 under both UV and visible light.[411] The visible-light absorption can be assigned to an electronic transition from the valence band to the doping energy level of the Sn^{4+} ions. This was located 0.4 eV below the conduction band and acted as an electron acceptor level. In contrast, Klosek et al. demonstrated that the visible absorption in V^{4+}-doped TiO_2 resulted in the

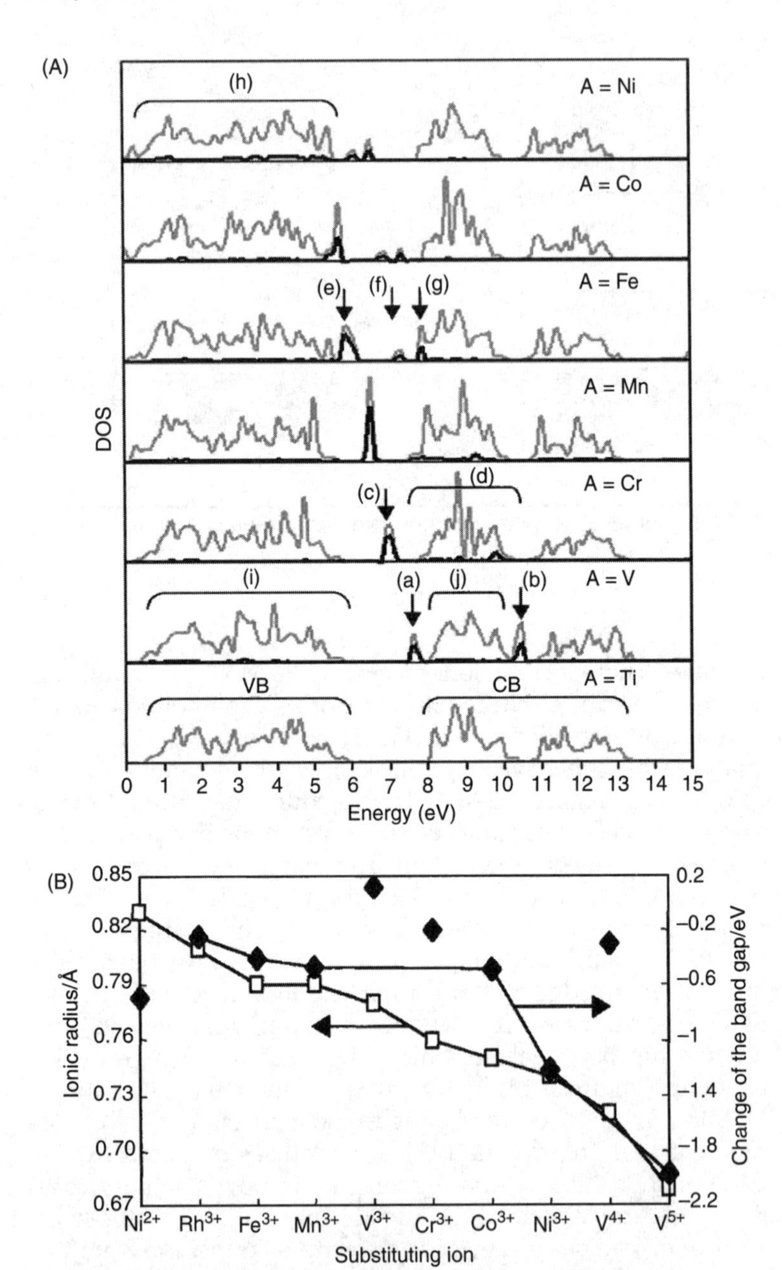

FIGURE 6.2 (A) DOS of the metal-doped TiO_2 ($Ti_{1-x}A_xO_2$: A = V, Cr, Mn, Fe, Co, or Ni). Gray solid lines: total DOS. Black solid lines: dopant's DOS. The states are labeled (a) to (j) (from Ref. 409, Copyright 2002, Elsevier). (B) The relationship between the ionic radius and the change of the bandgap (anatase). Except for some cations, the bandgaps decrease with decreasing the cation size. (From Ref. 410, Copyright 1999, The Chemical Society of Japan.)

photoexcitation from the V 3*d* electron donor level to the TiO_2 conduction band.[412] This brings in a more efficient visible-light-driven activity for ethanol photooxidation over V^{4+}-doped TiO_2 than that of pure TiO_2. The visible-light-driven photocatalytic activity of Fe^{3+}-doped TiO_2 for water splitting was also attributed to the photoexcitation of doping the donor level (i.e., Fe 3*d* orbitals) to the TiO_2 conduction band.[413–415]

Anpo and co-workers prepared various metal-ion-doped TiO_2 materials using advanced ion implantation.[416–421] They found that the absorption band of metal-ion-implanted TiO_2 (with the metals V, Cr, Mn, Fe, and Ni) shifted smoothly toward the visible light region. The extent of the red shift depended on the amount and type of metal ions being implanted, though the absorption maximum and minimum values stayed constant. Such a shift not only allowed the metal-ion-implanted TiO_2 to use solar irradiation more effectively, but also initiated effective photocatalytic reactions under both UV and visible-light irradiation. It was found that Pt^{4+} and Ag^{+}-doped TiO_2 nanoparticles also exhibited improved photocatalytic activities under visible-light or UV irradiation. This was explained as follows: The doping ions in these photocatalysts not only contributed to the visible-light absorption, but also served as an recombination inhibitor by trapping electrons or holes, which in turn promoted the charge separation required for the photocatalytic reaction.[422,423] However, in some cases, the metal-ion dopants could also serve as the recombination sites for photoinduced charges and not as a recombination inhibitor. This resulted in relatively low photocatalytic activity even under UV irradiation.

In 1994, Choi et al. investigated the effects of metal ion doping on the photocatalytic activity of TiO_2.[424] They found that the photocatalytic activity was related to the electron configuration of the dopant ion. Of the 21 metal-ion dopants studied, Fe, Mo, Ru, Os, Re, V, and Rh ions doping significantly increased the photocatalytic activity, whereas Co and Al ions doping caused detrimental effects. The nature of the metal ions doped in the TiO_2 significantly affected the charge recombination and electron transfer rates. An EPR study of doped TiO_2 colloids showed that Fe- or V-doped TiO_2 caused the growth of the Ti^{3+} signal. These changes were attributed to an inhibition of the hole-electron recombination by the Fe^{3+} or V^{4+} dopant. On the other hand, the Mo^{6+} dopant behaved as an irreversible electron trap.[425]

Kudo et al. reported that Ni^{2+}, Rh^{3+}, or Cr^{3+}-doped TiO_2 exhibited virtually no photocatalytic activity, whereas TiO_2 co-doped by Ni^{2+}, Rh^{3+}, or Cr^{3+} together with Ta^{5+}, Nb^{5+} or Sb^{5+} led to efficient O_2 evolution from water.[397,426–429] Reaction conditions required Ag^+ as an electron acceptor and visible-light irradiation. Doping by Ni^{2+}, Rh^{3+}, or Cr^{3+} created donor levels in the bandgap and made TiO_2 responsive to visible light. The charge balance was maintained by co-doping of Ta^{5+}, Nb^{5+}, or Sb^{5+}, which was needed to suppress the recombination between photogenerated electrons and holes. Ohno et al. expanded the effective wavelength of the TiO_2 photocatalyst into the visible region by Ru doping.[430] With this particular photocatalyst, the oxygen evolution reaction occurred under the irradiation of visible light at wavelengths longer than 440 nm. Iron (III) ions were used as the electron acceptor. Kahn et al. found that Pt, Ir, and Co-ionized titania nanotubes prepared by an ion-exchange method were effective photocatalysts for the production of stoichiometric hydrogen and oxygen by water splitting under visible-light irradiation.[431,432] Pt, Ir, and Co ionization all reduced the bandgap energy. Visible light-responsive TiO_2, obtained by self (Ti^{3+}) doping, was found to decompose water in to methanol or silver nitrate solution under visible-light irradiation. The observed absorption and photocatalytic ability in the visible-light region was ascribed to the defect levels present within the bandgap of these materials.[433–436]

Attention has also been paid to other oxide semiconductors as host photocatalysts for metal-ion doping. Cr-doped $SrTi_{1-x}Cr_xO_3$ (x = 0.00, 0.02, 0.05, 0.10) was prepared by a solvothermal method and showed increasing photocatalytic activities for hydrogen production both under UV and visible light with increasing amounts of chromium. The new bandgap in the visible-light range obtained by Cr doping was attributed to the band transition from the Cr 3*d* to the Cr 3*d* + Ti 3*d* hybrid orbital.[398] Wang et al. studied the photocatalytic properties of $SrTiO_2$ doped with Cr cations on different sites.[437] Interestingly, $(Sr_{0.95}Cr_{0.05})TiO_3$ with Cr^{3+} doped at the Sr^{2+} sites showed higher photocatalytic activity of H_2 evolution under visible light than $Sr(Ti_{0.95}Cr_{0.05})O_3$ with Cr cations (mixed Cr^{3+} and Cr^{6+}) doped at the Ti^{4+} sites. It was concluded that because the Cr^{6+} cations created the potential level of the empty Cr^{6+} lower than that for H_2 evolution, they behaved as the trapping center for photoinduced electrons. Thus, Cr^{6+} should be avoided in Cr-containing visible-light-driven photocatalysts designed for water splitting.

Kudo and co-workers found that Mn- and Ru-doped $SrTiO_3$ showed photocatalytic activities for O_2 evolution from an aqueous silver nitrate solution.[438] Ru-, Rh-, and Ir-doped $SrTiO_3$ loaded with Pt co-catalysts produced H_2 from an aqueous methanol solution under visible-light irradiation ($\lambda > 440$ nm). In particular, the Rh (1 percent)-doped $SrTiO_2$ gave a quantum yield of 5.2 percent at 420 nm for H_2 evolution. The visible-light response of Rh-doped $SrTiO_3$ was resulted from the transition from the electron donor level formed by the Rh ions to the conduction band composed of Ti 3*d* orbitals. When $SrTiO_3$ was co-doped with Sb^{5+}/Cr^{3+}, Ta^{5+}/Cr^{3+}, or Ta^{5+}/Ni^{2+} all the resulting compounds displayed efficient photocatalytic activities for hydrogen production from the aqueous methanol solutions under visible-light irradiation ($\lambda > 420$ nm).[397,427–439] When Ti^{4+} was replaced by Ni^{2+} or Cr^{3+} in $SrTiO_3$, the results were similar to those for Sb^{5+}/Cr^{3+} co-doped TiO_2. The charge became unbalanced and recombination centers formed. When a second metal ion such as Ta^{5+} or Sb^{5+} was co-doped to compensate for the charge unbalance and suppress the formation of the recombination centers, visible-light absorption was maintained.

Miyake et al.[440] and Chen et al.[441] demonstrated that both Rh^{3+}- and Cu^{2+}-doped $CaTiO_3$ were good candidates for visible-light-driven oxide photocatalysis for hydrogen evolution. Reaction occurred under visible-light irradiation with methanol as the electron donor. Miyake et al. also found that Rh^{3+}-doped $Ca_3Ti_2O_7$ was active in photocatalytic hydrogen generation under visible light.[442] Moreover, it was found that Rh^{3+}-doped $Ca_3Ti_2O_7$ with a layered structure showed higher stability in air than Rh^{3+}-doped $CaTiO_3$ with a simple perovskite structure. Lee et al. studied the effects of both Cr and Fe cation substitution on the photophysical and photocatalytic properties of the layered perovskite $La_2Ti_2O_7$.[399,443] The contribution of these dopants led to the formation of a partially filled 3*d* band, which served as the electron donor level in the bandgap of $La_2Ti_2O_7$. It also caused the excitation of electrons from this localized interband to the conduction band of $La_2Ti_2O_7$ and was responsible for visible-light absorption for the H_2 evolution from water under visible-light. Suzuki et al. reported that Fe^{3+} and W^{6+} substitution for Ti^{4+} in $K_2La_2Ti_3O_{10}$ resulted in a small activity for hydrogen production under visible-light.[444] Under the same condition, pure $K_2La_2Ti_3O_{10}$ showed no activity.

Recently, it was reported that both V-doped and Zn-doped $K_2La_2Ti_3O_{10}$ exhibited high photocatalytic activities of hydrogen production under visible light.[445] The hybridization of either V 3*d* or Zn 3*d* and O 2*p* electron orbitals resulted in a new localized energy level. The catalyst was easily excited with lower energy, which in turn improved the photoactivity of $K_2La_2Ti_3O_{10}$ for water splitting. The bandgap and photocatalytic activity of In_2TiO_5 also underwent significant changes as a result of V doping. These enabled the absorption of radiation from the entire visible region of 400 to 800 nm and led to the improvement of photocatalytic hydrogen production under visible light.[446] Zou et al. investigated the doping effects of different metal ions (Mn, Fe, Co, Ni, Cu) on the structural and photocatalytic properties of a $InTaO_4$ photocatalyst,[447–451] as shown in Figure 6.3. Of these, $In_{0.9}Ni_{0.1}TaO_4$ showed the highest photocatalytic activity, which induced direct splitting of water into stoichiometric amounts of oxygen and hydrogen under

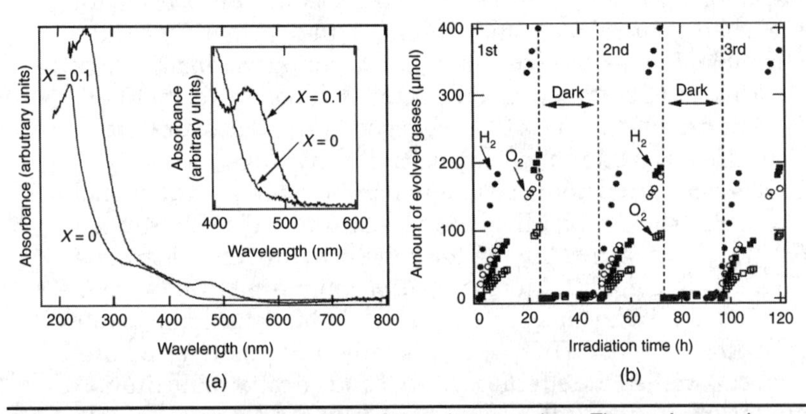

FIGURE 6.3 (a) Optical properties of the photocatalyst. The main panel shows the ultraviolet-visible diffuse reflectance spectra of $In_{1-x}Ni_xTaO_4$ (x = 0 and 0.1) at room temperature, with the inset providing an expanded view of the spectra in the wavelength region from 400 to 600 nm. (b) Photocatalytic H_2 and O_2 generation. Shown are the evolution of H_2 and O_2 from pure water using as catalyst a suspension of $NiO_y/In_{0.9}Ni_{0.1}TaO_4$ (solid circles, H_2; open circles, O_2) and $RuO_2/In_{0.9}Ni_{0.1}TaO_4$ (solid squares, H_2; open squares, O_2). Experiments were done using 0.5 g catalyst powder suspended in 250 mL pure water in a Pyrex glass cell under visible light (>420 nm). (From Ref. 447, Copyright 2001, Nature Publishing Group.)

visible-light irradiation with a quantum yield of about 0.66 percent at 402 nm. The narrower bandgap of Ni-doped $InTaO_4$ was attributed to the Ni 3*d* donor level formed in the forbidden band.[449] Ni-modified $K_4Nb_6O_{10}$ prepared by a solid-state reaction also exhibited increased visible-light absorption and photocatalytic activity for hydrogen evolution. This was believed to be the effect of the Ni^{2+} doping.[452] Cu 3*d* donor level formed above the valence band of $BiTaO_4$ by Cu^{2+} doping. This contributed to the increased photocatalytic activity for hydrogen evolution under visible light compared to Cu-$BiTaO_4$.[453] The strong photoabsorption and good performance of H_2 evolution in the visible-light region shown by Cr^{3+}-doped $Bi_4Ti_3O_{12}$ was largely attributed to the occurrence of the impurity level Cr 3*d* in both conduction and valence bands.[454]

Kudo et al. controlled the oxidation numbers of Ir when it substituted for Nb^{5+} and Ta^{5+} ions in the B sites of a perovskite structure for $NaBO_3$ (B = Nb, Ta).[455] Their strategy consisted of co-doped alkaline earth metal and lanthanum ions for Na^+ ions at the A sites. This contributed to maintaining the charge balance in $NaBO_3$. The resulting $NaNbO_3$:Ir/A (A: Sr, Ba, and La) showed H_2 or O_2 evolution, whereas $NaTaO_3$:Ir/A showed H_2 evolution under visible light. Yang et al. reported La^{3+}/Cr^{3+} co-doped $NaTaO_3$ showed intense visible-light absorption and H_2 production in the presence of methanol.[456] The photocatalytic activity was much higher than that of monodoped $NaTaO_3$. A high visible-light photocatalytic activity for O_2 evolution from an aqueous solution containing an electron acceptor (quantum yield of 6 percent at 420 nm) was found for Cr^{6+}-doped $PbMoO_4$. It should be noted that the formation of an electron-acceptor level of Cr 3*d* orbitals below conduction band as a result of the Cr^{6+}-replacement was believed to have given rise to the absorption bands and photocatalytic activity in the visible-light region.[457]

Although ZnS is a highly efficient photocatalyst for H_2 evolution because of its high conduction band level, it is only active in the UV light region.[458] Hence, it would be quite interesting to see if ZnS becomes responsive to visible light while maintaining its high H_2 evolution activity following designed modification. With this in view, a series of Cu- and Ni-doped ZnS photocatalysts were prepared. The results showed high activities for H_2 evolution from the aqueous solutions containing sulfite ions under visible-light irradiation (λ > 420 nm) even in the absence of a platinum co-catalyst. The visible-light response was a result of the formation of 3*d* donor levels by the doped Cu^{2+} and Ni^{2+} in the wide bandgap of the ZnS.[400,401,459]

In contrast to the doping modifications to create donor and/or acceptor bandgap states, which worked mostly UV-light-active photocatalysts, some visible-light-driven photocatalysts were also modified by metal-ion doping. The aim is to narrow their bandgaps in order to utilize visible light in even longer wavelength region to perform more efficient photocatalytic water splitting. Reber et al. modified the visible-light-active photocatalyst CdS by doping with Ag^+ to extend the spectral response up to 620 nm.[377] A significant enhancement in the photocatalytic activity for hydrogen evolution was obtained with CdS powders containing 0.77 percent Ag^+. Ren et al. reported two series of photocatalysts based on CdS with a little of $MInS_2$ (M = Cu, Ag) as the dopant.[460] The resulting compounds had narrower bandgaps and higher photocatalytic activities than undoped CdS. More recently, a large degree of Mn doping was discovered to be effective in improving the photocatalytic activity and stability of CdS for visible-light hydrogen production.[461]

Liu et al.[462] and Xu et al.[463–465] reported that the Cu-doped $Zn_xCd_{1-x}S$ solid-solution photocatalyst with the absorption edge shifted to a lower energy region displayed higher water-splitting efficiency to produce hydrogen than $Zn_xCd_{1-x}S$, which itself had been shown to be an excellent visible-light-driven photocatalyst.[466,467] Zhang et al. found Ni^{2+} doping also greatly improved the photocatalytic activity of $Zn_xCd_{1-x}S$ for hydrogen production.[468] It has been known that $ZnIn_2S_4$ can efficiently produce hydrogen from some aqueous solutions with sulfite and sulfide ions as the electron donor under visible light irradiation.[469,471] Shen et al. hydrothermally synthesized a series of Cu-doped $ZnIn_2S_4$ photocatalysts with broader absorptions in the visible-light region than the corresponding undoped $ZnIn_2S_4$.[472] As shown in Figure 6.4, they found the photocatalytic activity of $ZnIn_2S_4$ was remarkably enhanced by Cu doping. The 0.5 wt% Cu-doped $ZnIn_2S_4$ photocatalyst showed the highest activity for hydrogen evolution under visible-light irradiation. The surplus doped Cu^{2+} ions served as the recombination sites for the photogenerated electrons and holes.[473]

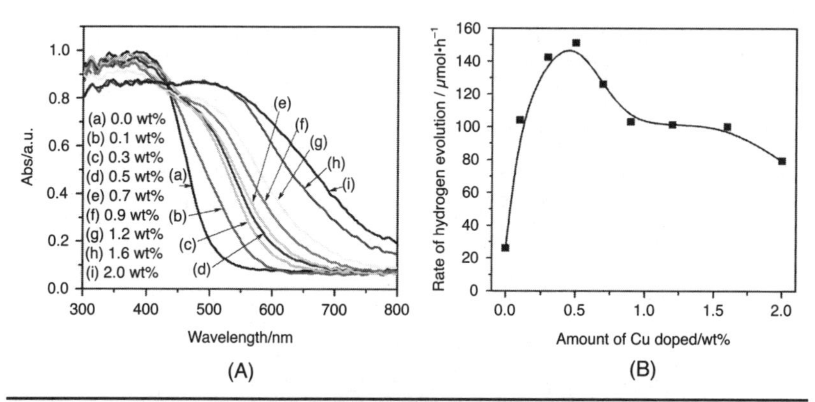

Figure 6.4 (A) Diffuse reflectance spectra of Cu-doped $ZnIn_2S_4$. (B) Dependence of photocatalytic activity for H_2 evolution over Cu-doped $ZnIn_2S_4$. The concentrations of Cu^{2+}were (a) 0.0 wt% (b) 0.1 wt%, (c) 0.3 wt%, (d) 0.5 wt%, (e) 0.7 wt%; (f) 0.9 wt%, (g) 1.2 wt%, (h) 1.6 wt%, (i) 2.0 wt%, respectively. (From Ref 472. Copyright 2008, American Chemical Society.)

6.1.2 Nonmetal Doped TiO_2

Nonmetal ion doping is another approach used to modify UV-light-active photocatalysts. It has been widely used to narrow the bandgap and improve the visible-light-driven photocatalytic activity. Unlike metal-ion dopants, nonmetal-ion dopants are less likely to form donor levels in the bandgap but instead shift the valence band edge upward. This results in a narrowing of the bandgap as indicated in Figure 6.5. The technology of nonmetal ion doping has been widely used to modify some UV-light-active oxide photocatalysts, such as Ti-based oxides,[474–481] Ta-based oxides,[482–489] Zr-based oxides and Nb-based oxides,[490–493] and so on.

Various nonmetal ions (such as C, N, S) were used to dope TiO_2, and the products were studied for their optical and photocatalytic properties. Nonmetal ion doped TiO_2, whose absorption spectra were red shifted to longer wavelengths, exhibited improved photocatalytic activities compared to those for pure TiO_2, especially in the visible-light region.[375,494–506] Chen et al. used XPS to show that additional electronic states exist above the valence band edge of pure TiO_2 for C-, N-, and S-doped TiO_2 via XPS (Figure 6.6a).[507,508] This additional electron density of states can explain the red-shifted absorption of these potential photocatalysts, as observed in the shoulder and tail-like features in the UV and Vis spectra (Figure 6.6b).

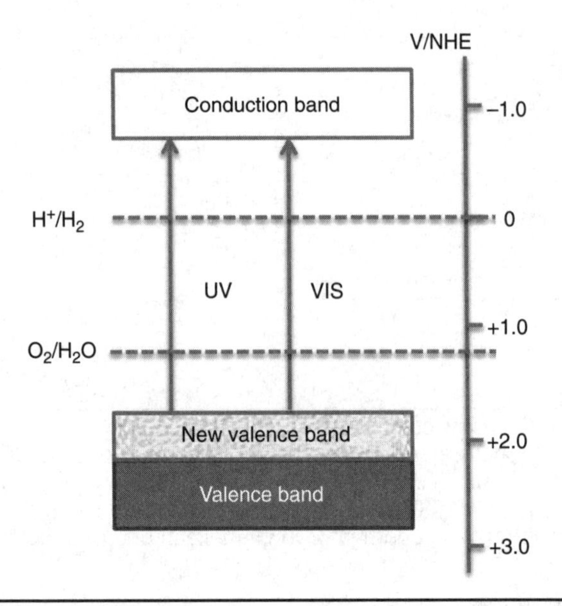

FIGURE 6.5 New valence band formation on the top of valence band by doping of nonmetal ions.

Asahi et al. studied the substitutional doping of C, N, F, P, and S for O in anatase TiO_2.[509] They calculated the electronic band structure of anatase TiO_2 with different substitutional dopants, as shown in Figure 6.7. It was found that the substitution of N for O, which led to the mixing of the 2*p* states of N with the 2*p* of O, was most effective. It led to the bandgap

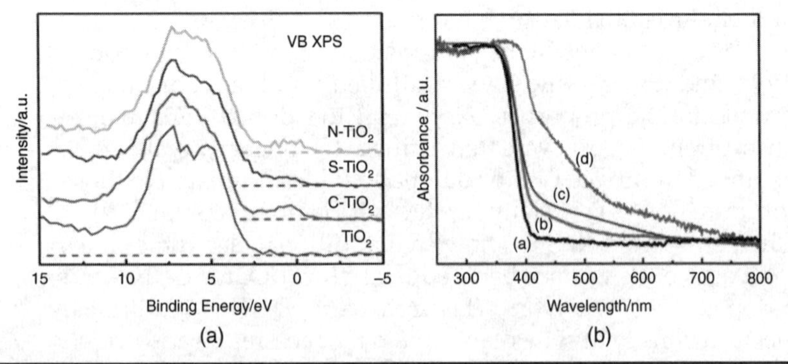

FIGURE 6.6 VB XPS spectra (A) and diffuse reflectance spectra (B) of (a) pure TiO_2, (b) C-TiO_2, (c) S-TiO_2, and (d) N-TiO_2. (From Ref. 508, Copyright 2008, American Chemical Society.)

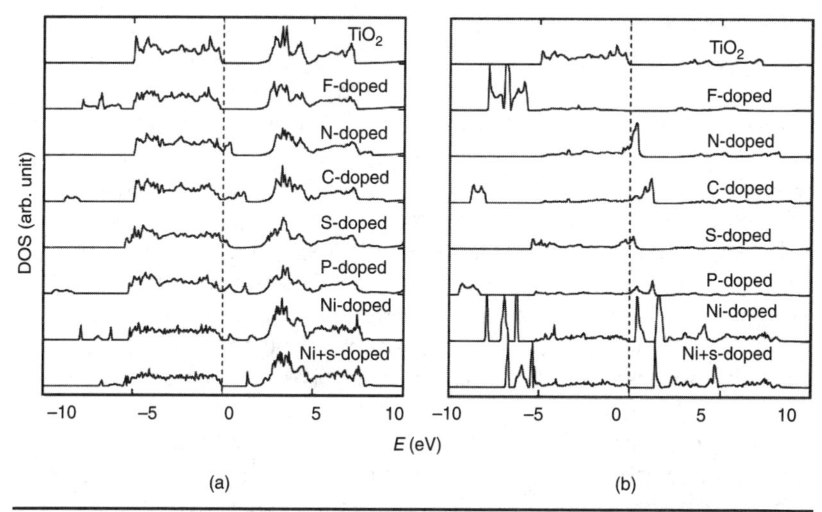

Figure 6.7 (a) Total DOS of doped TiO_2 and (b) the projected DOS into the doped anion sites, calculated by FLAPW. The dopants F, N, C, S, and P were located at a substitutional site for an O atom in the anatase TiO_2 crystal (the eight TiO_2 units per cell). The results for N doping at an interstitial site (N_i-doped) and at both substitutional and interstitial sites (Ni_{i+s}-doped) are also shown. (Reprinted with permission from Ref. 509, Copyright 2001, AAAS.)

narrowing by shifting the valence band edge upward, which in return resulted in N-doped TiO_2 having a much higher photocatalytic activity than pure TiO_2 in the visible light region. Chen et al. found that the degree of nitrogen doping was responsible for the significant increase in photocatalytic activity observed in the TiO_2 nanocolloid versus the nitrided commercial nanopowder.[504] Braun et al. revealed an additional e_g resonance in the valence band of TiO_2 formed by nitrogen doping in the oxygen 1s NEXAFS pre-edge.[510] This extra resonance was found to bear co-responsibility for the photocatalytic performance of N-doped TiO_2 at visible-light wavelengths.

N-doped TiO_2 has been both widely investigated and successfully prepared by many different methods. These include the physical/chemical vapor deposition,[511,512] heating of titanium hydroxide and urea,[513] reactive magnetron sputtering,[514,515] hydrothermal/solvothermal process,[516,517] and treating TiO_2 powder in an NH_3 gas flow.[518,519] Yuan et al. synthesized N-doped TiO_2 by heating a mixture of urea and TiO_2 at 350–700°C in air.[520] XPS confirmed that both the chemisorbed N_2 and the substitutional N contributed to the material's response to visible light. The substitutional N was considered the predominate factor in improving the photocatalytic activity of the water splitting under visible light. Lin et al. demonstrated that the N-doped TiO_2 photocatalyst, synthesized by a two-microemulsion technique, showed favorable photocatalytic hydrogen evolution at neutral pH range in methanol/water solutions.[521] Pillai and co-workers reported that the chemical modification of titanium isopropoxide using different nonmetallic chemical reagents such as urea, sulfuric acid, and trifluoroacetic acid improved both the anatase stability and the photocatalytic activity by doping with nitrogen, sulfur, and fluorine, respectively.[522–524]

The photocatalytic activity of S-doped TiO_2 has also been studied in depth. Using different synthetic methods, anionic sulfur can be doped into TiO_2 to replace the lattice oxygen or as a cation to replace the Ti ion.[497,525–528] It was found that S-doped TiO_2 showed different photocatalytic activities under visible light.[529] In Nishijima and co-workers' study of S-doped TiO_2, a higher activity level for hydrogen evolution than that for N-doped TiO_2 photocatalysts under visible light was reported.[530] Kahn et al. prepared C-doped TiO_2 with main rutile structure by pyrolyzing Ti metal in a natural gas flame. The C-doped TiO_2 possessed lower bandgap energy and displayed a much higher photoactivity in water splitting than pure TiO_2 with the mixed phase of antanse and rutile.[531] It was found that C-doped TiO_2 nanotubes also displayed a high photoactivity for water splitting.[532] The bandgap reduction and the new intragap band formation in C-doped TiO_2 nanotubes extended its absorption of solar energy up to the visible and infrared region.[532]

To further enhance the visible-light activity, non-metal co-doping of TiO_2, such as F/B-co-doping,[533,534] F/N-co-doping,[535,536] S/F-co-doping,[537] F/C-co-doping,[538] C/N-co-doping,[539,540] S/N-co-doping,[541,542] N/Br-co-doping,[543] B/N-co-doping,[544,545] N/Si-co-doping,[546] C/S-co-doping,[547] and P/F-co-doping,[548] have all been studied. Results suggested that nonmetal co-doped TiO_2 compounds had significant visible photocatalytic activities due to the synergetic effect of the co-doping. Luo et al. found that Cl/Br co-doped TiO_2 displayed a much higher photocatalytic activity for water splitting than either Cl- or Br-doped TiO_2.[549] Domen et al. found that N/F co-doped TiO_2 had high visible-light photocatalytic activity for water oxidation owing to the absorption band of $TiN_xO_yF_z$ in the visible region.[550–552] N/S co-doped TiO_2 showed considerable activity in the photocatalytic water splitting. This has been attributed to the visible-light photoexcitation of the electrons from the isolated energy levels in the bandgap formed by the doped N or S to the conduction band.[553] Liu et al. synthesized B/N co-doped TiO_2 with high visible-light photocatalytic activity.[554] They proposed that the synergistic effects of B/N co-doping created visible-light absorption by N doping and the lifetime of the photoinduced carriers was prolonged by B doping. Subsequently, Li et al. found that the visible-light-driven photocatalytic activity of B/N co-doped TiO_2 for hydrogen evolution increased greatly compared to that of N doped TiO_2.[555] They believed that the co-doping of boron contributed to an excellent activity of B/N co-doped TiO_2, because the doped boron could act as shallow traps for electrons, which prolong the life of photoinduced electrons and holes. Recently, OCN-doped TiO_2 nanoparticles were prepared for photocatalysis in the visible-light region of 380 to 550 nm as indicated by UV-vis absorption spectra.[466] In this doped TiO_2, the OCN group bonding to titanium atoms produce a weakening of its C=O double bond and a hardening of CN and NO bonds, which enables visible absorption and related photocatalytic activity.[556]

As with N-doped TiO_2, N-doping in Ta_2O_5 caused a valence band shift upward to a higher potential energy, which narrowed the bandgap.[485,557] TaON and Ta_3N_5, prepared from reacting Ta_2O_5 with NH_3, were found to be active for water splitting into hydrogen and oxygen under visible-light irradiation.[484,558–562] In particular, in the presence of the sacrificial electron acceptor (Ag^+), TaON functioned as a very efficient photocatalyst for the oxidation of water into O_2 (quantum yield of 34 percent).[560] Recently, nonmetal-ions-doped titanates and tantalates have also been studied for photocatalytic reduction and oxidation under visible-light irradiation. Wang et al. found that in the visible-light range and near ultraviolet range, N-doped and N/S co-doped $SrTiO_3$ displayed much higher photocatalytic activities for NO elimination than pure $SrTiO_3$.[563,564] Kasahara et al. found that under visible-light irradiation $LaTiO_2N$ reduced H^+ into H_2 and oxidized H_2O into O_2 in the presence of a sacrificial electron donor (methanol) or acceptor (Ag^+) using the bandgap transition (2.1 eV).[565,566] Moreover,

$LaTiO_2N$ could split water without any sacrificial reagent under visible-light irradiation when used as a photoanode in the photoelectrochemical cell.[567] The visible-light absorption was attributed to the new valence band composed of O 2*p* orbitals for the lower-energy side and N 2*p* orbitals for the higher-energy side. Similarly, due to N-doping reducing the bandgap energy, N-doped $KTa_{0.92}Zr_{0.08}O_3$ was reported to achieve a complete splitting of water under visible-light irradiation when Pt was loaded as a co-catalyst. Yamasita et al. prepared $MTaO_2N$ (M: Ca, Sr, Ba) by nitriding $M_2Ta_2O_7$ at 1123 K for 15 h.[568] $MTaO_2N$ had small bandgap energies (2.5 to 2.0 eV) and could absorb visible light at 500 to 630 nm via the N 2*p* orbitals of the upper regions of the valence bands.[489] Under visible-light irradiation, $MTaO_2N$ reduced H^+ into H_2 in the presence of a sacrificial electron donor. These oxynitrides, however, did not function in the oxidation of water due to the fact that the valence band did not have sufficient over potentials for the oxidation potential of water. Liu et al. reported $Y_2Ta_2O_5N_2$ as a novel photocatalyst with high activity for water splitting under visible-light irradiation in the presence of appropriate sacrificial reagents.[393] The smaller bandgap energy of $Y_2Ta_2O_5N_2$ was attributed to the partial replacement of O^{2-} by N^{3-} in $YTaO_4$ and the predominant population of the valence band by a hybrid orbital between N 2*p* and O 2*p*.

In addition to Ti- and Ta-based oxides as the host photocatalysts, some other oxides with wide bandgaps were modified by using nonmetal ion doping and were found to improve visible-light-driven photocatalytic activities for water splitting. Ji et al. studied photocatalytic water splitting using nitrogen-doped $Sr_2Nb_2O_7$ under visible-light irradiation.[491] After nitrogen doping, the $Sr_2Nb_2O_7$ bandgap energy was reduced, and subsequently, induced photocatalytic activity for hydrogen production from methanol-water mixtures in the visible-light region. N-doped ZrO_2, In_2O_3, C-doped In_2O_3, and Nb_2O_5 were also reported to show enhanced photocatalytic activities for water splitting under visible-light compared to the undoped oxide hosts.[569–572] Hisatomi et al. found that the novel spinel-type d^0d^{10} complex oxynitride photocatalyst $Zn_xTiO_yN_z$ displayed photocatalytic activity for both the reduction and oxidation of water in the presence of either a sacrificial electron donor or acceptor under visible light.[573]

Li et al. demonstrated the development of a simple lamellar-solid acid photocatalyst. N-doped HNb_3O_8 displayed superior visible-light-responsive photocatalytic activity in comparison to N-doped Nb_2O_5 and TiO_2. Kanade et al. reported that N-doped $Nb_2Zr_6O_{17}$ gave a quantum yield of 13.5 percent for photocatalytic hydrogen production from the decomposition of hydrogen sulfide in KOH aqueous solution under visible-light irradiation.[490] N-doped Ga-Zn mixed oxides with bandgap of 2.6 eV were capable of producing hydrogen from a methanol aqueous solution with an apparent quantum efficiency of 5.1 percent under visible-light illumination[574] with Rh/Cr_2O_3 was loaded as the co-catalyst.

Some other S_2 substituted metal oxides have also received considerable attention as visible-light-responsive photocatalysts for water reduction and/or oxidation. For example, the absorption edge of ZrW_2O_8 was significantly extended to the visible-light region by S doping.[575] H_2 and O_2 were evolved under irradiation of up to 360 nm and 510 nm, respectively. The visible-light sensitization was attributed to the S 3*p* states, which increased the width of the valence band itself and caused a decrease in the bandgap energy. Ishikawa et al. were the first to report a Ti-based oxysulfide, $Sm_2Ti_2S_2O_5$, as a visible-light-driven photocatalyst for hydrogen or oxygen production from aqueous solutions containing a sacrificial electron donor or acceptor.[576–578] It was found that the S 3*p* orbitals constituted the upper part of the valence band and that these orbitals made an essential contribution to the small bandgap energy. Subsequently, it was proposed that the members of the series $Ln_2Ti_2S_2O_5$ (Ln = Pr, Nd, Sm, Gd, Tb, Dy, Ho, and Er) would also function as visible-light-responsive photocatalysts for water splitting. The photocatalytic activity of [Pr, Nd, Er]$_2Ti_2S_2O_5$, containing sulfur defects and Ti as Ti^{3+} was lower than that of the other $Ln_2Ti_2S_2O_5$ forms. It appeared that the band structure was strongly affected by the lanthanoid ions.[579]

A number of La-based oxysulfides were shown to catalyze H^+ reduction to form H_2 and/or water oxidation to form O_2 under visible irradiation in the presence of a sacrificial electron donor (methanol, Na_2S-Na_2SO_3) and acceptor (Ag^+), respectively. The valence bands of these photocatalysts involved overlap of the O 2*p* and S 3*p* orbitals, where the higher-energy region mainly consisted of S 3*p* orbitals and the lower-energy region was mainly composed of O 2*p* orbitals.[580–582] Thus, it was deduced that the valence bands of photocatalysts consisting of S 3*p* orbitals, instead of O 2*p* orbitals, will result in the formation of narrow bandgaps. From the point of view of total substitution of O^{2-} by S^{2-} in oxide semiconductors, some novel ternary sulfide systems, including Zn(Cu)-In-S,[469,470,583–586] Ag-Ga(In)-S,[587,588] Cu-Ga-S(Se),[589,590] and Na-In-S,[591] have been developed as visible-light-driven photocatalysts for hydrogen production from aqueous solution containing a sacrificial reagent, such as Na_2S/Na_2SO_3.

6.2 Metal/Nonmetal Ion Co-doping

Metal/nonmetal ion co-doped semiconductor systems have been employed as photocatalysts in order to improve their photocatalytic activity under visible-light irradiation. The modification of TiO_2 by co-doping with metal and nonmetal ions was frequently employed to improve the photocatalytic activity, especially for pollutant degradation. For instance, (Ce + C, I, N or B),[592–595] (Fe + N or C),[596,597] (Bi + S, C or N),[598,599] (Ni + B or N),[600,601] (La + N, I or S),[602–604] and (Eu, Ta, Mo, Pt or W + N)[605–609] co-doped TiO_2, without exception, exhibited enhanced visible-light photocatalytic activity compared to their undoped counterparts. $SrTiO_3$ co-doped with La and N was also reported as having greater visible-light photocatalytic activity than the unmodified $SrTiO_3$.[610–612] However, only a few studies reported metal/nonmetal ion co-doped photocatalyst systems with modified conduction and valence band used for visible-light hydrogen production from water. Gai et al. proposed to dope TiO_2 using charge-compensated donor-acceptor pairs such as (N + V), (Nb + N), (Cr + C), and (Mo + C).[613] Among all these systems, TiO_2:(Mo + C) has the highest positive effect on photocatalytic water splitting. This is because it reduces the bandgap to the ideal visible-light region. It does not, however, have much effect on the conduction band position. Using density-functional theory calculations, Yin et al. predicted that (Mo, 2N) and (W, 2N) were the best donor-acceptor combinations in the low-alloy concentration regime, whereas (Nb, N) and (Ta, N) were the best choice of donor-acceptor pairs in the high-alloy concentration regime for solar-driven photoelectrochemical water-splitting.[614]

Sasikala et al. found co-doping of TiO_2 with In and N in cationic and anionic sites, respectively, resulted in the narrowing of the bandgap compared to In or N doping alone.[615] As a result, the In and N co-doped samples exhibited enhanced absorption of visible light and improved photocatalytic activity for hydrogen production. Due to the charge-compensation effect from the donor-acceptor co-doping (N, Al), co-doped ZnO obtained by the RF magnetron sputtering method displayed significantly reduced bandgap and enhanced photocurrents under visible-light irradiation, when compared to ZnO or Al-doped ZnO.[616] Kudo et al. synthesized Pb and halogen co-doped ZnS as an active visible-light-driven photocatalyst for H_2 evolution without co-catalysts.[617] Pb doping was responsible for the visible-light absorption, and halogen doping suppressed the formation of nonradiative transition sites in which recombination of photogenerated electrons and holes could occur.

Lei et al. synthesized sulfur-substituted and Zn-doped $In(OH)_3$ in an aqueous solution of ethylenediamine using the hydrothermal method and investigated its photoactivity for H_2 production under visible-light irradiation.[618] As shown in Figure 6.8, the bandgap of $In(OH)_3$ was narrowed by the

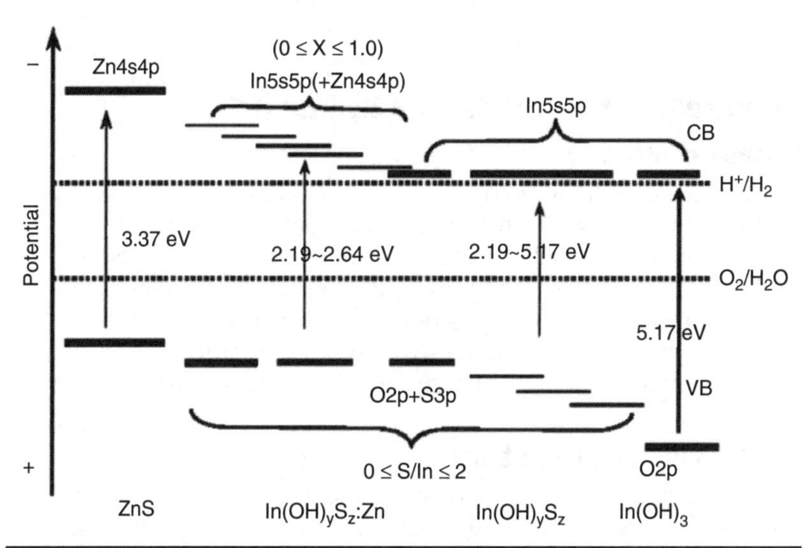

FIGURE 6.8 The proposed band structure of $In(OH)_3$ $In(OH)_yS_z$, and $In(OH)_yS_z$:Zn. (From Ref. 618, Copyright 2006, Elsevier.)

substitution of S^{2-} for OH^-. The valence band was composed mainly of S 3*p* orbitals hybridized with O 2*p* orbitals. Meanwhile, Zn^{2+} doping leveled up the conduction band consisting of In 5*s*5*p* and Zn 4*s*4*p* orbitals, and provided a large thermodynamic driving force for the reduction of water to produce H_2.

6.3 Doped ZnO

ZnO is a wide bandgap semiconductor, which has attracted considerable attention due to its potential technological applications, such as high efficient vacuum fluorescent displays (VFD) and field emission displays (FED).[619,620] ZnO has also been used for short wavelength laser devices,[621,622] high power and high frequency electronic devices,[623] and light-emitting diodes (LED).[624,625] ZnO shows many advantages: (1) it has a larger exciton energy (60 meV) than GaN (23 meV), which is useful for efficient laser uv applications; (2) the bandgap is tunable from 2.8 to 3.3 eV and from 3.3 to 4 eV in the alloys with Cd and Mg, respectively;[626,627] (3) wet chemical synthesis is possible; (4) it has low-power threshold at room temperature; and (5) dilute Mn-doped ZnO shows room-temperature ferromagnetism.[628]

6.4 Controlling Band Structure by Making Solid Solutions

In addition to using foreign elements for doping, forming solid solutions between wide and narrow bandgap semiconductors is another promising method for controlling photocatalyst band structure. Both the bandgap and position can be adjusted by varying the ratio of the compositions of the narrow and the wide bandgap semiconductor in the solid solution. Figure 6.9 shows the controllable band formation by making a solid solution.

6.4.1 Oxide Solid Solutions

The photophysical and photocatalytic properties of oxide semiconductors with similar crystal structures were also studied in order to explore the possibility of energy structure control using solid solutions.[243,299,629–631] Kudo et al. reported the

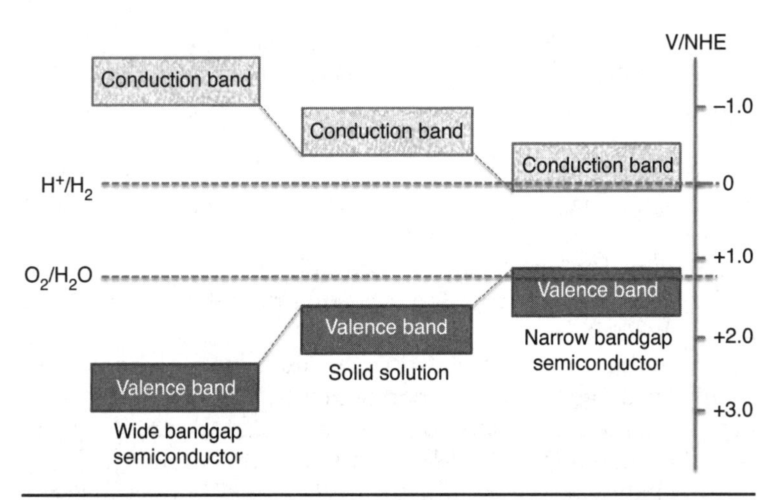

FIGURE 6.9 Band structure and bandgap change controlled by making a solid solution.

successive changes in absorption, photoluminescence spectra, and conduction band levels of In_2O_3-Ga_2O_3 solid solutions.[299] Moreover, they found that both the bandgap and conduction band level of the $Sr_2(Ta_{1-x}Nb_x)_2O_7$ solid solution with a layered perovskite structure could be controlled by changing the ratio of niobium to tantalum. All the Sr_2 $(Ta_{1-x}Nb_x)_2O_7$ photocatalysts studied had bandgaps larger than 3.9 eV and only exhibited photocatalytic activities for water splitting into H_2 and O_2 under UV irradiation.[249] Meanwhile, Zou et al. prepared $BiTa_{1-x}Nb_xO_4$ ($0 < x < 1$) solid solution photocatalysts using solid-state reactions.[632–633] These produced hydrogen both from aqueous CH_3OH/H_2O solutions and from pure water under UV irradiation. Although none of these photocatalysts exhibited visible-light-driven activities, these studies provided some important information for developing novel visible-light-driven photocatalysts derived from oxide solid solutions that could be used for water splitting.

Luan et al. prepared $Bi_xIn_{1-x}TaO_4$ ($0 < x < 1$) solid solutions using a solid-state reaction.[634] The bandgaps of the $Bi_xIn_{1-x}TaO_4$ ($x = 0.2$, 0.5, and 0.8) photocatalysts were estimated at about 2.86, 2.71, and 2.74 eV, respectively. It was suggested that the band structure consisted of a conduction band of mainly Ta 5*d*, In 5*p*, and In 5*s* orbitals, while the valence band was principally

O 2*p*, Bi 6*s*, and Bi 6*p* orbitals. Under visible-light irradiation ($\lambda > 420$ nm), H_2 and O_2 were evolved from CH_3OH and $AgNO_3$ aqueous solutions, respectively, using the $Bi_xIn_{1-x}TaO_4$ photocatalysts. Yi et al. prepared $Na_{1-x}La_xTa_{1-x}Co_xO_3$ monophase solid solutions by conventional solid-state reactions.[635] The $Na_{0.9}La_{0.1}Ta_{0.9}Co_{0.1}O_3$ photocatalyst exhibited the highest performance of H_2 evolution (4.34 μmol/h) under visible-light irradiation. The photocatalytic activities of $Na_{1-x}La_xTa_{1-x}Co_xO_3$ series were mainly attributed to the hybridization of the Co 3*d* and O 2*p* orbitals.

Since the divalent metals in scheelite-type molybdates are easily exchangeable with other elements, Yao et al. developed new scheelite solid-solution photocatalysts by combining the scheelite molybdates ($CaMoO_4$ and $(Na_{0.5}Bi_{0.5})MoO_4$) with the narrow bandgap semiconductor $BiVO_4$.[636,637] They crystallized with similar structures. Both the resulting $CaMoO_4$-$BiVO_4$ and $(Na_{0.5}Bi_{0.5})MoO_4$-$BiVO_4$ solid solutions showed high activities for photocatalytic O_2 evolution under visible-light irradiation. In fact, they were even better than that of monoclinic $BiVO_4$, which itself is a well-known efficient visible-light-driven photocatalyst. In Li's study, another series of novel solid-solution photocatalysts $Na(Bi_xTa_{1-x})O_3$ were successfully prepared using a simple hydrothermal method, and $Na(Bi_{0.08}Ta_{0.92})O_3$ showed the highest photocatalytic activity for hydrogen production under visible light.[638] This was attributed to the band structure formed by a hybrid conduction band of the (Bi 6*s* + 6*p* +Ta 5*d*) orbitals.

Wang et al. developed a novel series of solid-solution semiconductors, for example, $(AgNbO_3)_{1-x}(SrTiO_3)_x$ $(0 < x < 1)$ as visible-light-active photocatalysts for efficient O_2 evolution and decomposition of organic pollutants.[639,640] The modulation of the band structure (bandgap energy, band edge positions, etc.) depends on the extent of the orbital hybridization between both the Ag 4*d* and O 2*p* orbitals, and between the Nb 4*d* and Ti 3*d* orbitals. As a result of competition between the absorption ability to visible-light and the reductive/oxidative abilities, the highest visible-light activities for both O_2 evolution and decomposition of gaseous 2-propanol were realized over $(AgNbO_3)_{0.75}(SrTiO_3)_{0.25}$. Between solid solutions, of Y_2WO_6 and Bi_2WO_6, $BiYWO_6$ was found by Liu et al. to have the photocatalytic ability to stoichiometrically split water into H_2 and O_2 ratios under visible light up to $\lambda < 470$ nm.[641] It was suggested that the Bi 6*s* and Y 4*d* orbitals contributed to a new valence band and conduction band, respectively. It was expected that, due to the flexible structure of the solid solution, the activity would be favorably promoted by changing the ratio of Y_2WO_6 and Bi_2WO_6. Soon after, they discovered another $Bi_{0.5}Dy_{0.5}VO_4$ solid solution composed of $BiVO_4$ and $DyVO_4$ that responded to visible light up to 450 nm and completely split water into H_2 and O_2.[642]

6.4.2 Oxynitride Solid Solutions

Starting from the colored oxynitride $LaTiO_2N$, new perovskite-type solid solutions $LaTiO_2N$-$ATiO_3$ (A = Sr, Ba) were prepared by the thermal ammonolysis method.[643,644] Narrowed bandgaps were achieved by both lowering the bottom of the conduction band and raising the top of the valence band as x increased. Thus, $(SrTiO_3)_{1-x}(LaTiO_2N)_x$ presented suitable band positions for photocatalytic water splitting into hydrogen and oxygen under visible-light irradiation.

Since 2005, Domen's group has systematically investigated GaN-ZnO solid solutions as potentially new efficient photocatalysts capable of decomposing water into hydrogen and oxygen stoichiometrically under visible-light irradiation.[645–653] The solid solution of GaN and ZnO, $(Ga_{1-x}Zn_x)(N_{1-x}O_x)$, should have bandgaps greater than 3 eV because of the large bandgap energies of both GaN and ZnO (>3 eV). However, the *p-d* repulsion between the N 2*p* and Zn 3*d* orbitals shifted the valence-band maximum upward without affecting the conduction-band minimum. This resulted in a narrowing of the bandgap of GaN-ZnO solid solution as schematically depicted in Figure 6.10.[650,654] The result, therefore, was that the band positions of $(Ga_{1-x}Zn_x)(N_{1-x}O_x)$ were suitable for overall water splitting under visible-light irradiation. The quantum efficiency at 420 to 440 nm was about 2.5 percent, when $Rh_{2-y}Cr_yO_3$ was loaded as a co-catalyst.[646] The photocatalytic performance of $Rh_{2-y}Cr_yO_3/(Ga_{1-x}Zn_x)(N_{1-x}O_x)$ was due to the fact that the charge recombination was prevented and there was enhanced reactivity of photoexcited holes in the O_2 evolution reaction.[655] Moreover, the visible-light-driven photocatalytic activity of $Rh_{2-y}Cr_yO_2/(Ga_{1-x}Zn_x)(N_{1-x}O_x)$ was further improved by a post calcination treatment through reduction of the density of the zinc- and/or oxygen-related defects that functioned as recombination centers for

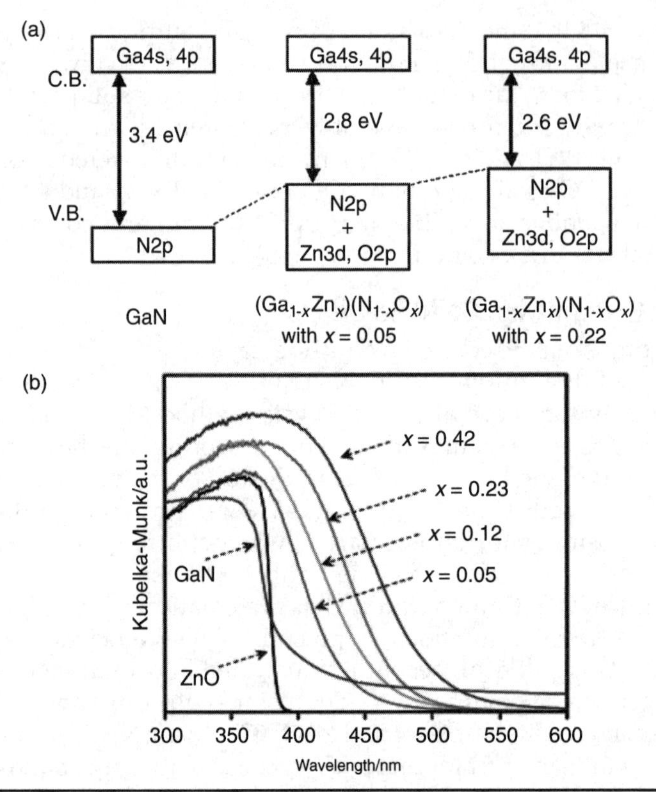

FIGURE 6.10 (A) Schematic band structures (from Ref. 650, Copyright 2005, American Chemical Society). (B) UV-visible diffuse reflectance spectra of $(Ga_{1-x}Zn_x)(N_{1-x}O_x)$. (From Ref. 51, Copyright 2007, American Chemical Society.)

photogenerated electrons and holes in the material. The maximum quantum efficiency obtained by post-calcination treatment is ca. 5.9 percent at 420 to 440 nm.[656] This was about an order of magnitude higher than the photocatalytic activity of the previously discussed photocatalysts used in overall water splitting under visible light (Ni-$InTaO_4$: 0.66 percent at 402 nm;[447] $BiYWO_6$: 0.17 percent at 420 nm).[641] In addition, an appropriate amount of electrolyte (e.g., NaCl and Na_2SO_4) in the photocatalytic reactant solution proved quite effective in enhancing overall water splitting using $Rh_{2-y}Cr_yO_3$-loaded $(Ga_{1-x}Zn_x)(N_{1-x}O_x)$.[657]

Based on the development of the $(Ga_{1-x}Zn_x)(N_{1-x}O_x)$ solid solution, the solid solution of $ZnGeN_2$ and ZnO: $(Zn_{1+x}Ge)(N_2O_x)$[658–661] was developed by Domen's group and found to be another active and stable photocatalyst for overall water splitting under visible-light irradiation. Similar to the $(Ga_{1-x}Zn_x)(N_{1-x}O_x)$ solid solution, the bandgap narrowing of $(Zn_{1+x}Ge)(N_2O_x)$ was also attributed to the *p-d* repulsion between the N 2*p* and Zn 3*d* orbitals. Thus, the visible-light-driven overall water splitting catalyzed by $(Zn_{1+x}Ge)(N_2O_x)$ proceeded via bandgap photoexcitation from the valence band formed by the N 2*p*, O 2*p*, and Zn 3*d* orbitals to the conduction band consisting of Ge 4*s* and 4*p* hybridized orbitals. Instead of the successful formation of the ideal solid solution between $(Ga_{1-x}Zn_x)(N_{1-x}O_x)$ and InN, the obtained Ga-Zn-In mixed oxynitride also behaved as a visible-light-driven photocatalyst for water splitting under visible-light irradiation, but it also photocatalyzed H_2 and O_2 evolution in the presence of an appropriate electron donor or acceptor.[662]

6.5 Organic and Inorganic Systems

Fundamental questions in chemistry and materials science still remain. What are the intermediate states of chemical reactions? How do we apply advanced spectroscopic tools to prove a mechanistic pathway? With detailed knowledge of the microscopic environment in which chemical reactions take place, can we engineer better catalysts?

An approach to developing artificial photosynthetic assemblies for the production of alcohols is based on all inorganic photocatalytic units arranged on a mesoporous support. The principal challenge is to design units that complete the two necessary half-reactions—the oxidation of water and the reduction of carbon dioxide efficiently in the visible range, rather than the UV light range. The components consist of molecular or polynuclear multielectron catalysts (either for water oxidation or carbon dioxide reduction) coupled to a visible-light charge transfer pump that acts as the chromophore in the artificial photosynthetic system. Oxobridged heterobinuclear units of the type TiOCr, TiOMn, TiOCo, and TiOCe with broad metal-to-metal charge-transfer transitions (MMCT) in the visible region (350–600 nm) serve as these charge separators. When visible light comes in, an electron is transferred from one metal to the other, creating a metal with a high oxidizing potential (i.e., Mn(III) or Cr(IV)) and one with a reducing potential Ti(III). The polynuclear multielectron catalysts on the side of water oxidation are transition metal oxide nanoclusters, based on IrO_2, Co_3O_4, and Mn_2O_3. All components are anchored on a silica mesoporous support (MCM-41) that allows for the arranging and coupling of sites, a large surface area for catalysis, and the separation of the reduction and oxidation half-reactions.

CHAPTER 7

Surface and Morphology

7.1 Surface and Interface Chemistry

Due to their scientific and technological importance, inorganic single crystals with highly reactive surfaces have long been studied. Unfortunately, surfaces with high reactivity usually diminish rapidly during the crystal growth process as a result of the minimization of surface energy.

The electron-hole pair formation that occurs at the interface between a semiconductor and a solution upon absorption of light leads to oxidation or reduction of solution species. The fabrication of artificial photosynthetic systems for the conversion of H_2O and CO_2 to fuels (for example, H_2 and CH_3OH) has become a field of research interest lately and has encouraged new fundamental investigations of the interfacial interactions of light, electron flow, and chemical reactions.

The photophysical mechanisms of photocatalysts are not well understood. Key questions include the following:

1. Are the transition metal ions located primarily on the surface or in the lattice?
2. Is the surficial binding of substrates affected by doping?
3. Do transition metal ions influence charge-pair recombination?
4. Do altered interfacial transfer rate constants associated with surficial transition metal ion complexes play a primary role in altered photochemical kinetics?

In recent years, considerable effort has been devoted to the study of Fe(III)-doped titanium dioxide in order to improve the photocatalytic efficiency of TiO_2.[663–666] The generally accepted mechanism to explain this improved photocatalytic performance is the formation of shallow charge trapping sites within the TiO_2 matrix as well as on the particle's surface through the replacement of Ti(IV) by Fe(III) ions.[665] Thus, the undesirable recombination of electron/hole pairs generated upon ultra-bandgap irradiation can be partially prevented.

The correlations between the surface and the electronic structure of nanoparticles (NPs) and their catalytic properties will provide information on which surface is the most selective and active for a given catalyst. The activation energies obtained from this study will shed light on how the NPs evolve and whether the NP surface is stable in a given environment and how fast the surface changes.

The surface stability and reactivity of inorganic single crystals have been thought to be dominated by their surface chemistry, whose effect on the equilibrium morphology is critical for the synthesis of single crystals with high reactivity.[331–333,667–677] For anatase TiO_2, both theoretical and experimental studies found that the minority (001) facets in the equilibrium state are especially reactive. However, large high-quality anatase single crystals with a high percentage of (001) facets have not been synthesized.[678–680] An early study[679] showed that the hydrothermal treatment of hydrous titanium(IV) oxide in the presence of hydrofluoric acid resulted in irregular aggregates of polymorphic TiO_2 with anhedral morphology. Recently, anatase single crystals were synthesized using chemical transport reactions, but the process had a long reaction time and the crystals were of low purity and had no (001) facets.[680] Therefore, preparation of uniform, high purity anatase single crystals with controllable crystallographic facets still remains a challenge.

Attempts have been made with various adsorbate atoms to change the relative stabilities of different crystal facets.[336,337,677–679] For anatase TiO_2, among oxygenated surfaces, (100) facets are the most stable, whereas under clean and hydrogenated conditions, (101) facets are the most stable.[671,675,677] However, both H- and O-terminated anatase surfaces present high surface energies, which restrict the formation of large anatase single crystals. High values of surface energy are mainly attributed to the high bonding energies of H-H (436.0 kJ mol^{-1}) and O-O (498.4 kJ mol^{-1}).[681]

To further explore the effects of various adsorbate atoms, Yang et al. carried out a systematic investigation of 12 nonmetallic atoms *X* (where *X* can represent H, B, C, N, O, F, Si, P, S, Cl, Br, or I) based on first-principle calculations.[682] The well-defined, high-purity anatase single crystals synthesized were very useful as model for fundamental studies in surface science. Furthermore, high-purity anatase single crystals with a high percentage of reactive (001) facets have promising applications in solar cells, photonic and optoelectronic devices, sensors, and photocatalysis. This result illustrates the power of combining first-principle calculations and experimental techniques to achieve engineering of surface and crystallographic characteristics of crystalline materials.

7.2 Nanostructure and Morphology

Advances in the synthesis of particles of nanometer dimensions, narrow size distribution, and controlled shape have generated interest because of the potential to create novel materials with tailored physical and chemical properties.[683,684] New properties arise from quantum confinement effects and from the increasing fraction of surface atoms with unique bonding and geometrical configurations. For example, cobalt nanocrystals display a wealth of size-dependent structural, magnetic, electronic, and catalytic properties.

Particle size, shape, surface composition, and surface structure are key factors controlling catalytic performance of nanoparticles. However, the properties of NPs are usually difficult to characterize at the atomic level under practical catalytic conditions, especially when the size of NPs is below 1 nm.[685] The ability to control the particle size and morphology of nanoparticles is of crucial importance nowadays in the applications of solar photovoltaic, sunlight water splitting and photoelectrochemical cells, chemical and biosensors, and so on. Quantum size effects on the exciton and bandgap energies were observed in semiconductor nanocrystals.[686,689,690] If the observed valence band (VB) or conduction band (CB) shifts are due to quantum confinement, one would expect the size of the band shifts to increase as the particle size of the nanocrystals is decreased.

Most nanomaterials are synthesized under nonequilibrium and tend to phase-segregate and to be strained. Therefore, the phase diagrams of NPs are different from those of the bulk materials. Studies of NP catalysts have shown that they can go through some profound changes under reaction conditions.[691–694] A fundamental understanding of the growth and properties of nanocrystals would greatly benefit from the detailed information of their electronic structure as a function of size and of the presence and nature of the molecules bound to their surface. Because the metal nanocrystals are extremely reactive and oxidize easily, it is important to use techniques that can interrogate the particles in their growth environment so that their electronic and chemical structure can be followed during growth and catalytic reactions.

The classical model for the growth of monodisperse nanocrystals assumes a discrete nucleation stage followed by slow growth via monomer attachment.[695,696] In addition, a relatively high monomer concentration in the resolution is considered to be necessary, and particle interactions should be avoided. However, many studies have suggested differently. Chen et al.[697] reported that the MnO colloidal nanocrystal system monomers in the precursor solution were depleted during particle size distribution focusing. Particle growth without the presence of precursor in solution was considered via an Ostwald ripening process that dissolved smaller particles allow larger particles to grow. Violations of the classical growth model have also observed in the nanocrystal growth of colloidal platinum.[698] It was found that the particle coalescences are highly involved during the growth. However, monodisperse nanocrystals can still be achieved in spite of the strong particle interactions. In order to better understand nanocrystal growth mechanism, it is critical to monitor changes of monomeric species during the growth in addition to the nanocrystal morphology.

7.2.1 Quantum-Sized Transition Metal Oxides

When the crystallite dimension of a semiconductor particle falls below a critical radius of approximately 10 nm, the charge carriers appear to behave quantum mechanically[699–710] as a simple particle in a box. As a result of this confinement, the bandgap increases and the band edges shift to yield larger redox potentials. The solvent reorganization free energy for charge transfer to a substrate, however, remains unchanged. The increased driving force and the unchanged solvent reorganization free energy in size-quantized systems are expected to increase the rate constant of charge transfer in the normal Marcus region.[711–713] Thus, the use of size-quantized semiconductor TiO_2 particles may result in increased photoefficiencies for systems in which the rate-limiting step is charge transfer.[714,715]

The use of size-quantized semiconductors[714–718] to increase photoefficiencies is demonstrated by several studies. However, in other work, size-quantized semiconductors have been found to be less photoactive than their bulk-phase counterparts.[715] In the latter cases, surface speciation and surface defect density appear to control photoreactivity.[719,720] The positive effects of increased overpotentials (i.e., the difference between E_{vb} and E_{redox}) on quantum yields can be offset by unfavorable surface speciation and surface defects due to the preparation method of size-quantized semiconductor particles.

7.2.2 TiO_2 Quantum Dots

The bandgap of TiO_2 nanoparticles will increase when the sizes are reduced at the level of quantum confinement (see Figure 7.1).[721] Two critical issues, however, need to be addressed to enable TiO_2-based materials to work effectively. First, the activity in the visible or near-IR spectral region must be enhanced. Second, charge carrier separation upon photoexcitation needs to be facilitated so these charges can participate in useful photochemistry instead of simply recombining. Semiconductor QDs, because of their unique and desirable properties, have attracted significant attention as possible sensitizers can drastically enhance the performance of TiO_2-based photovoltaic cells and photocatalysts. By selecting an appropriate semiconductor material and controlling the size of the QDs, a band alignment can be achieved between the QDs and TiO_2. The small bandgap of the QDs coupled with the band alignment favors visible or near-IR light excitation followed by electron transfer to TiO_2.

Different metals have been doped into TiO_2 nanomaterials.[402,403,425,722–748,750–752] The preparation methods of nonmetal-doped TiO_2 nanomaterials can be divided into three types: wet chemistry, high temperature treatment, and ion implantation on TiO_2 nanomaterials. Wet chemistry methods usually involve hydrolysis of a titanium precursor in a mixture of water and other reagents, followed by heating. Choi et al. performed a systematic study of TiO_2

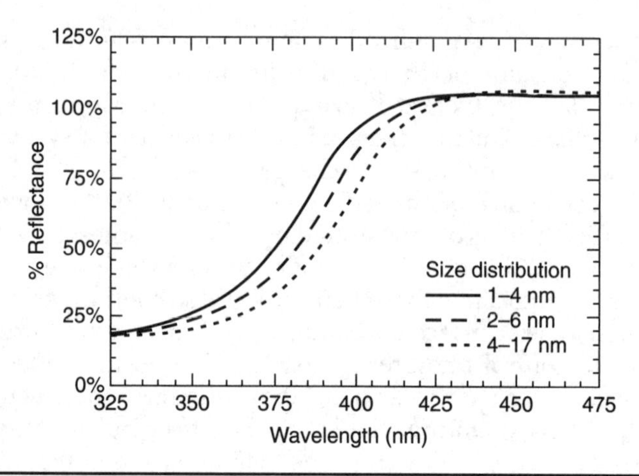

FIGURE 7.1 UV/vis reflectance spectra of size-quantized TiO_2 (from Ref. 721).

nanoparticles doped with 21 metal ions by the sol-gel method and found the presence of metal ion dopants significantly influenced the photoreactivity, charge carrier recombination rates, and interfacial electron transfer rates.[725] Li et al. developed La^{3+}-doped TiO_2 by the sol-gel process and found that the lanthanum doping could inhibit the phase transformation of TiO_2, enhance the thermal stability of the TiO_2, reduce the crystallite size, and increase the Ti^{3+} content on the surface.[731] Nagaveni et al. prepared W, V, Ce, Zr, Fe, and Cu ion-doped anatase TiO_2 nanoparticles by a solution combustion method and found that the solid solution formation was limited to a narrow range of concentrations of the dopant ions.[735] Wang et al. prepared Nd^{3+}-doped and Fe(III)-doped TiO_2 nanoparticles with a hydrothermal method and found that anatase, brookite, and a trace of hematite coexisted at lower pH (1.8 and 3.6) when the Fe(III) content was as low as 0.5 percent and the distribution of iron ions was nonuniform between particles, but at higher pH(6.0), the uniform solid solution of iron-titanium oxide formed.[747,750]

Anpo et al. prepared TiO_2 nanoparticles doped with Cr and V ions with an ion-implantation method.[416–421] Bessekhouad et al. investigated alkaline (Li, Na, K)-doped TiO_2 nanoparticles prepared by sol-gel and impregnation technology and found that the crystallinity level of the products was largely dependent on both the nature and the concentration of the alkaline, with the best crystallinity obtained for Li-doped TiO_2 and the lowest for K-doped TiO_2.[723] Cao et al. prepared Sn^{4+}-doped TiO_2 nanoparticle films by the plasma-enhanced CVD method and found that, after doping by Sn, more surface defects were present on the surface.[411] Gracia et al. synthesized M (Cr, V, Fe, Co)-doped TiO_2 by ion beam induced CVD and found that TiO_2 crystallized into the anatase or rutile structures depending on the type and amount of cations present with partial segregation of the cations in the form of M_2O_n after annealing.[729] Wang et al. synthesized Fe(III)-doped TiO_2 nanoparticles using oxidative pyrolysis of liquid-feed organometallic precursors in a radiation frequency (RF) thermal plasma and found that the formation of rutile was strongly promoted with iron doping compared to the anatase phase being prevalent in the undoped TiO_2.[753]

Choi et al.[754–756] have shown that selectively doped quantum- (Q-) sized particles have a much greater photoreactivity as measured by their quantum efficiencies for oxidation and reduction than their undoped counterparts. They present the results of a systematic study of the effects of 21 different metal ion dopants on the photochemical reactivity of quantum-sized TiO_2 with respect to both chloroform oxidation and carbon tetrachloride reduction. Enhanced photoactivity was seen for Fe(III), Mo(V), Ru(III), Os(III), Re(V), V(IV), and Rh(III) substitution for Ti(IV) at the 0.5 atom % level in the TiO_2 matrix. The maximum enhancements were 18-fold (CCl_4 reduction) and 15-fold ($CHCl_3$ oxidation) increases in quantum efficiency for Fe(III)-doped Q-TiO_2.

Choi et al.[754] used laser flash photolysis measurements to show that the lifetime of the blue electron in the Fe(III)-, V(IV)-, Mo(V)-, and Ru(III)-doped samples was increased to 50 ms, while the measured lifetimes of the blue electron in undoped Q-particles were <200 μs.[754–756] They established that the experimental quantum efficiencies for oxidation and for reduction could be correlated to the measured transient absorption signals of the charge carriers. In general, a relative increase in the concentration of the long-lived (ms) charge carriers results in a corresponding increase in photoreactivity. However, if an electron is trapped in a deep trapping site, it will have a longer lifetime but it may also have a lower redox potential that could result in a decrease in photoreactivity.

Reactivity of doped TiO_2 appears to be a complex function of the dopant concentration, the energy level of the dopants within the TiO_2 lattice, their *d*-electronic configuration, the distribution of dopants, the electron-donor concentration, and the light intensity.

7.2.3 Bandgap Engineering for Visible Light Response

Suitable band engineering is required in order to develop new photocatalysts for water splitting under visible-light irradiation as shown in Figure 7.2. In general, the conduction bands of stable oxide semiconductor photocatalysts are composed of empty orbitals (LUMOs) of metal cations with d^0 and d^{10} configurations. Although the valence band level depends on crystal structure and bond character between metal and oxygen, the level of the valence band consisting of O 2*p* orbitals is usually ca. 3.0 eV.[757] Accordingly, a new valence band or an electron donor level (DL) must be formed with orbitals of elements other than O 2*p* to make the bandgap (BG) or the energy gap (EG) narrower because the conduction band level should not be lowered. Not only the thermodynamic potential but also kinetic ability for 4-electron oxidation of water are required for the newly formed valence band.

The electron donor level is created above a valence band by doping some elements into conventional photocatalysts

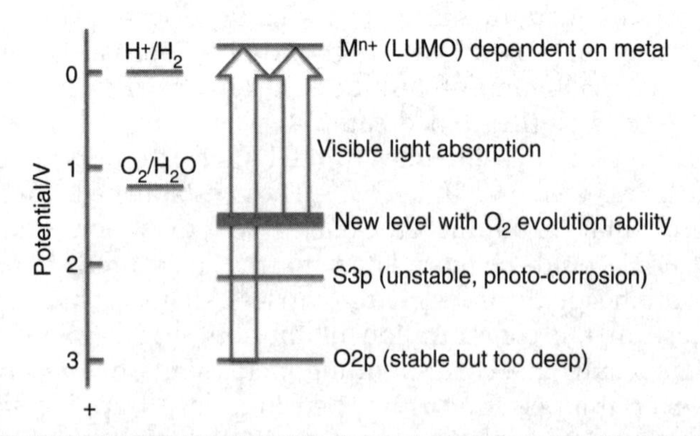

Figure 7.2 Band structure control to develop visible-light-driven-photocatalysts for water splitting.

with wide bandgaps such as TiO_2 and $SrTiO_3$. It results in the formation of an energy gap. On the other hand, some metal cations and anions can contribute to valence band formations above the valence band consisting of O 2*p* orbitals. Here, bandgap is distinguished from energy gap. The energy gap is formed by the impurity level that does not form a complete band. Making a solid solution is also a useful band engineering procedure.

WO_3 is one of the most well-known photocatalysts with visible-light response for O_2 evolution in the presence of sacrificial reagents such as Ag^+ and Fe^{3+}. Abe and co-workers found that Pt/WO_3 is active for degradation of acetic acid, CH_3CHO and IPA under visible light irradiation.[758]

Bi_2WO_6 and Bi_2MoO_6 with the Aurivillius structure are active for an O_2 evolution reaction under visible-light irradiation. These tungstate and molybdate photocatalysts are not active for H_2 evolution because of the low conduction band level. These photocatalysts are also used for degradation of HCHO,[759] CH_3OH,[760] CH_3COOH,[761–762] rhodamine B, and methylene blue.[763–775]

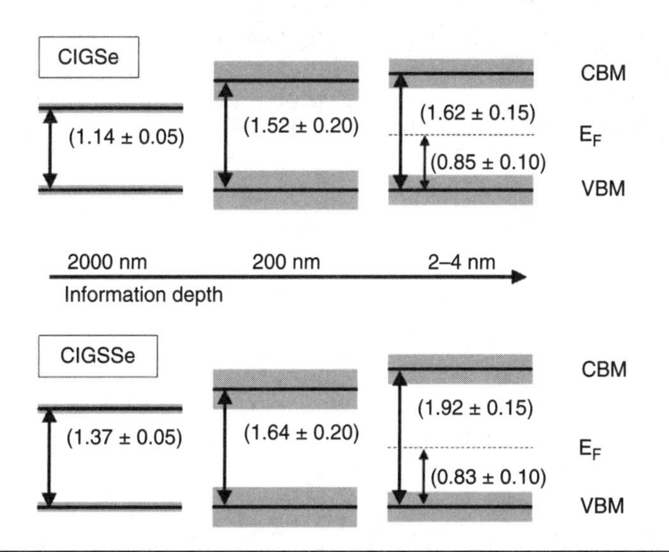

FIGURE 7.3 Schematic representation of the determined CIGSe (top) and CIGSSe (bottom) bandgap energies using the three different spectroscopic approaches (from Ref. 389).

7.2.4 Bandgap at the Surface

Bandgap and band edge positions, as well as the overall band structure of semiconductors, are of crucial importance for the photocatalysts. The energy position of the band edge can be controlled by the electronegativity of the dopants, solution pH (for example, flat-band potential variation of 60 mV per pH unit), and by quantum confinement effects. Accordingly, the band edges and bandgap can be tailored to achieve specific electronic, optical, or photocatalytic properties in nanostructure semiconductors.

Figure 7.3 indicates the formation of a surface region with significantly higher E_g. This E_g-widened surface region extends further into the bulk of the sulfur-free CIGSe thin film compared to the CIGSSe thin film.

7.2.5 Bandgap Change from Crystal Structure

Anatase and rutile are obtained through phase transition. The bandgap of anatase is 3.2 eV, while that of rutile is 3.0 eV, indicating that the crystal structure determines the bandgap even if the composition is the same. The difference in the bandgap between anatase and rutile is mainly due to the difference in the conduction band level. The conduction band level of anatase is higher than that of rutile, leading the difference in photocatalytic abilities between anatase and rutile (brookite TiO_2 is selectively prepared by a hydrothermal method).[776]

CHAPTER 8

Soft X-ray Spectroscopy and Electronic Structure

While utilizing solar energy will require new materials developed through the control of atomic, chemical, and electronic structures, achieving such control of properties requires an intimate collaboration between materials synthesis and characterization of the electronic properties of complex materials. To meet the challenge of characterization and interpretation, new advanced tools must be designed, built, and brought to bear on materials that can exhibit the novel properties required under a real-world environment.

So far, most of the developments in renewable energy materials have been achieved by the well-tested Edisonian method of trial and error. However, progress has been slow when compared to the rapid advances in other electronic devices, exemplified by Moores law for device density in silicon microchips or the increase in magnetic data storage density. This situation calls for a new strategy, where slow evolution of traditional concepts is accelerated by feedback from spectroscopy. A close feedback loop between synthesis, analysis, and prediction enables a more rational design of new materials than trial and error. For example, by tailoring the electronic energy levels of the absorber molecules, the donor/acceptor molecules for electron-hole separation, and the eventual transport to a conducting electrode, one can minimize the energy loss in a solar cell. These energy levels can be determined by incisive, element-specific spectroscopic techniques based on synchrotron radiation.

Synchrotron radiation-based soft x-ray spectroscopy has become a powerful tool to determine the bandgap properties of semiconductors.[687,688] X-ray originates from an electronic transition between a localized core state and a valence state. Soft x-ray absorption spectroscopy (XAS) probes the local unoccupied electronic structure (conduction band); soft x-ray emission spectroscopy (XES) probes the occupied electronic structure (valence band); and the addition of resonant inelastic soft x-ray scattering (resonant Raman spectroscopy with soft x-rays) can tell the energy levels that reflect the chemical and physical properties of semiconductors. Recently, quantum size effects on the exciton and bandgap energies were observed in single-walled carbon nanotubes (SWNTs).[777]

In particular, it was possible to apply the detailed knowledge obtained in soft x-ray spectroscopy of bulk rutile and anatase TiO_2 and hematite α-Fe_2O_3 to the study of the interaction of adsorbates on undoped and doped TiO_2 and α-Fe_2O_3 surfaces. This is potentially important, for instance, for photolysis in relation to both solar cells and to the photocatalytic production of hydrogen.

However, it is difficult to establish a direct correlation between the surface composition on bimetallic NPs and the observed catalytic activity, since soft x-ray spectroscopic characterizations performed under the corresponding reaction conditions for the activity measurements are still a challenge. Various photon-in/photon-out techniques are now developed to permit in situ characterization of NPs during catalytic reactions under atmospheric pressures. As the size of NPs decreases below 1 nm, XES and XAS techniques in fluorescence detection, which are usually referred to as techniques for bulk materials, become surface specific, since most atoms in these small NPs are surface atoms.

8.1 Soft X-ray Absorption and Emission Spectroscopy

Soft x-ray absorption spectrum provides information about the unoccupied states. For example, in oxygen K-edge absorption, the oxygen 1*s* electron is excited to empty electronic states in the carbon allotrope conduction band, and the dipole selection rule provides a tool to study locally the C 2*p* character of these unoccupied valence bands (Figure 8.1).

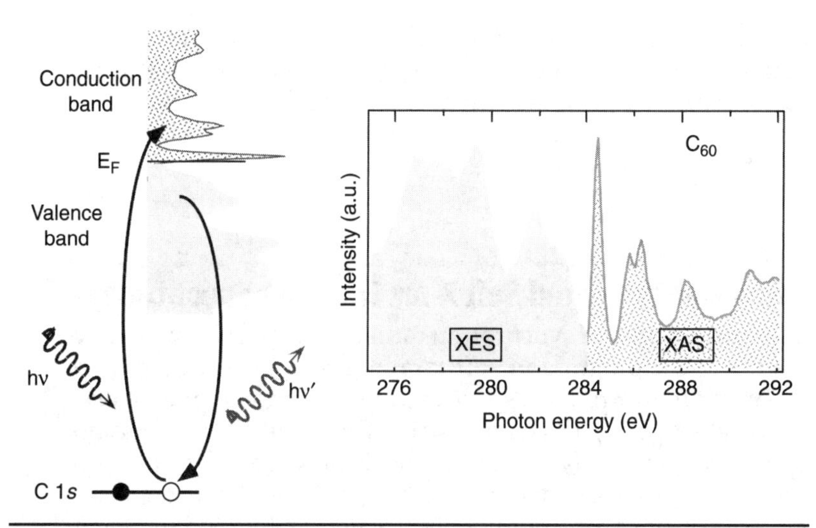

FIGURE 8.1 The schematic illustration of x-ray absorption and emission processes and XAS-XES spectra.

The atomic nature of the core hole implies elemental and site selectivity. The probability of such a transition is related to the x-ray absorption cross-section. The intensity of these secondary electrons or the photons can be measured as a function of incoming photon energy. This will reflect the absorption cross-section as the intensity of the secondary electrons/emitted photons is proportional to the absorbed intensity. Because of the short mean free path of electrons, the electron-yield-detection method is surface sensitive (about 1–5 nm). If the out coming photons are detected (fluorescence yield), the x-ray absorption becomes bulk sensitive (about 100–200 nm) due to the comparatively larger attenuation length of photons. Figure 8.1 gives an example of XAS study of C_{60}, and the XAS spectrum (at the right side) reflects the density of states (DOS) of conduction band.

The core vacancy left by the excited 1*s* electron is filled by an electron from the valence orbital; thus, soft x-ray emission gives direct information about the chemical bonding. Figure 8.1 also gives an example of XES study of C_{60} and the XES spectrum (at the left side) reflects the density of states (DOS) of the valence band. In addition to the inherent elemental selectivity of x-ray spectra, energy selective excitation allows separation of features that pertain to different atoms of a sample. Resonant excited emission from chemically nonequivalent sites of the same atomic species can be separated. The interest in the technique is presently booming due to the advent of third-generation synchrotron radiation sources.

8.2 Resonantly Excited Soft X-ray Emission Spectroscopy

The introduction of synchrotron radiation did not immediately lead to great progress in soft x-ray emission spectroscopy in the way other core-level spectroscopy, like photoemission and x-ray absorption, developed. The first soft x-ray emission spectroscopic study using monochromatized synchrotron radiation was carried out in 1987.[778] X-ray absorption and emission have been traditionally treated as two independent processes, with the absorption and emission spectra providing information on the unoccupied and occupied electronic states, respectively. The formulations of resonant inelastic x-ray scattering (RIXS) led to a Kramers-Heisenberg type dispersion formula for the cross-section with generally only the resonant part of the scattering process taken into account.[779] Second-order perturbation theory for the RIXS process leads to the Kramers-Heisenberg formula for the resonant x-ray scattering amplitude. Using this starting point RIXS has been analyzed in periodic solids as a momentum conserving process, suggesting that it can be used as a novel band-mapping technique.[780] The same starting point was adopted to unravel the symmetry selective properties of RIXS in theoretical works focused on molecules.[781–783] Resonant inelastic x-ray scattering at core resonances has become a new tool for probing the optical transitions in transition metal oxides.[784–786] Final states probed via such a channel, RIXS, are related to the eigenvalues of the ground state Hamiltonian. The core-hole lifetime is not a limit on the resolution in the RIXS process. According to the many-body picture, an energy of a photon, scattered on a certain low-energy excitation, should change by the same amount as a change in an excitation energy of the incident beam, so that inelastic scattering structures have constant energy losses and follow the elastic peak on the emitted-photon energy scale. Figure 8.2 shows an example that such an energy loss is originated from the *dd* excitations observed at the Ti L-edge.

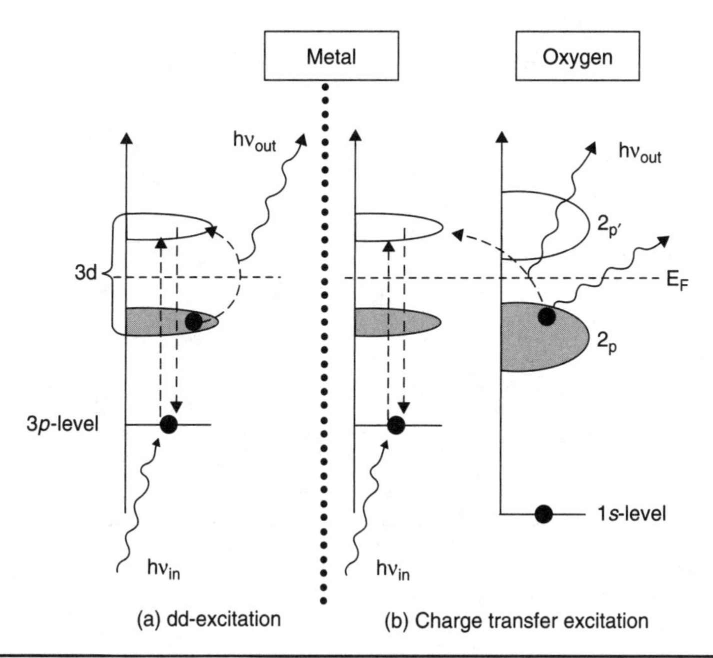

FIGURE 8.2 The schematic illustration of resonant inelastic x-ray scattering (RIXS) in which d-d and charge transfer excitations can be probed.

8.3 Electronic Structure of Metal Oxide Catalysts

Soft x-ray absorption and emission spectroscopy is used to determine the bandgap of semiconductors.[787,788] Quantum size effects on the exciton and bandgap energies were observed in single-walled carbon nanotubes (SWNTs) using soft x-ray spectroscopy.[789]

8.3.1 TiO_2

The research interest in nanostructured TiO_2 is based on the possibilities of using the material in various applications, such as dye sensitized solar cells,[790] displays,[791] and Li-ion batteries.[792]

Figure 8.3 shows the Ti 2*p* XAS spectra of nanoporous TiO_2. The x-ray absorption spectrum is derived from the two L_3 ($2p_{3/2}$) (457–462 eV) and L_2 ($2p_{1/2}$) (462–467 eV) parts,

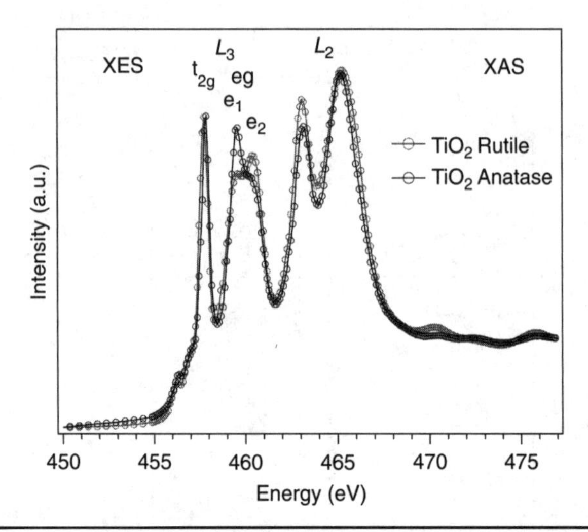

FIGURE 8.3 Ti 2p x-ray absorption spectra of TiO_2 in crystal structure of anatase and rutile.

further split into sharp t_{2g} and doublet split e_g (e_1 and e_2) states due to slight distortion from the O_h symmetry. The e_1 peak at the lower energy side originates from the long Ti-O bonds due to a hybridization effect that is weaker than the short Ti-O bonds. Note that the intensity ratio of the doublet split e_g is reversed in rutile TiO_2 because of the slightly different crystal symmetry (D_{2h}) in comparison with anatase TiO_2 under the D_{2d} crystal field.[48] Below the L_3 threshold two well-separated peaks (in 456–457 eV) are observed in the absorption spectrum. These have a predominantly triplet character and are mixed through the spin orbit interaction and the Coulomb repulsion into the main L_3 edge.[49] For the XES spectra of both rutile and anatase TiO_2 (see in Figure 8.4), the main band with peak maximum centered around 451 eV corresponds to the transition from the peak of valence density of states to the Ti $2p_{3/2}$ core state; lower intensity structures are seen at the higher energy side (at 457.5 eV) due to the Coster-Kronig process that transfers the hole from Ti $2p_{1/2}$ to Ti $2p3/2$ and to a multiple electron excitation.

O K-edge x-ray absorption and emission spectra of both rutile and anatase TiO_2 recorded at total electron yield detection modes are depicted in Figure 8.5. The well-defined pre-edge is

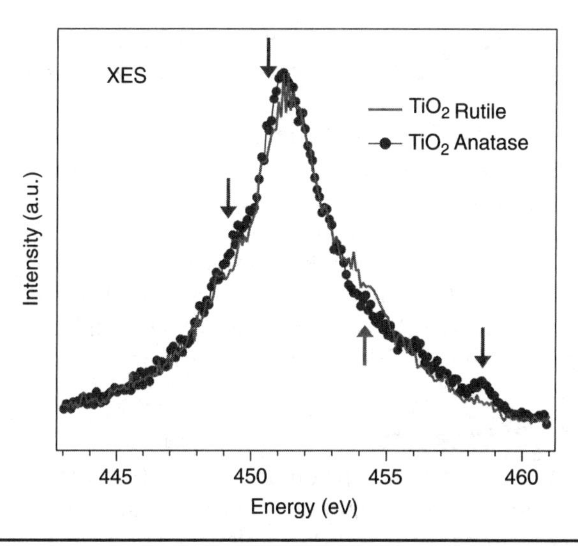

FIGURE 8.4 Ti 2p x-ray emission spectra of TiO_2 in crystal structure of anatase and rutile.

attributed to O 2*p* weighting of states that have a predominantly transition-metal 3*d* character, that is, Ti-3*d* O-2*p* mixing. The 3*d* states are split into two bands, which are assigned to t_{2g} and e_g symmetries, although this assignment is not strictly valid, as the crystal field is slightly distorted from octahedral symmetry.

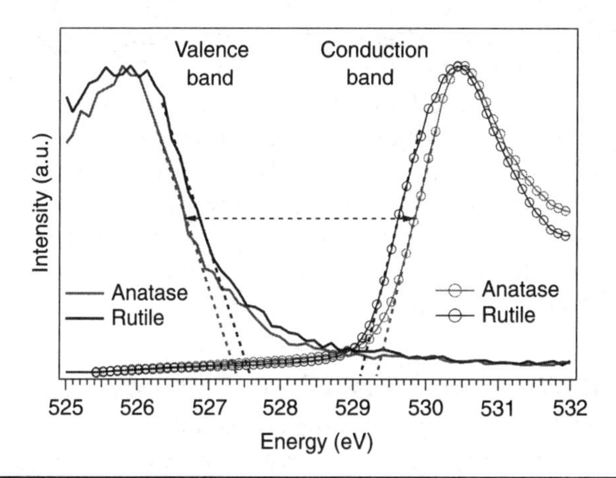

FIGURE 8.5 O 1s x-ray absorption and K-edge x-ray emission spectra of TiO_2 in crystal structure of rutile (dark) and anatase (grey).

The valence-core XES spectra of the rutile and anatase TiO_2 aligned together with the corresponding XAS spectra are shown. The bandgap is determined to be 1.9 for rutile and 2.4 eV for anatase TiO_2, respectively (note: it will obtain larger bandgap for anatase and rutile TiO_2 if one uses the first derivatives of XAS and XES to determine the conduction band minimum (CBM) and valence band maximum (VBM). In the case of anatase, the absorption-emission spectrum shows a larger bandgap. The enlarged bandgap in anatase TiO_2 is seen as the moving down of the valence band maximum and moving up of the conduction band minimum.

8.3.2 Fe_2O_3 Catalysts

Although the bandgap of hematite (α-Fe_2O_3), reported to be around 1.9 to 2.2 eV (depending on its crystalline status and methods of preparation), and its valence band edge are suitable for oxygen evolution, the conduction band edge of hematite is too low to generate hydrogen. Therefore, a blue shift of the bandgap of hematite of about 0.3 to 0.6 eV and the concomitant upward shift of the conduction band edge would make hematite an ideal anode material for photocatalytic devices for the photo-oxidation of water in terms of the cost, abundance, and nontoxicity, as well as thermal and structural stability and photocorrosion resistance.

Figure 8.6 shows the investigation of quantum confinement effects on bandgap profiling in α-Fe_2O_3 arrays by resonant inelastic soft x-ray scattering for potential application of such nanomaterials in direct photo-oxidation of water by solar irradiation. The 2.5-eV excitation,[793] which corresponds to the bandgap transition of hematite, appears significantly enlarged compared to the reported 1.9 to 2.2 eV bandgap of single crystal and polycrystalline samples. The finding strongly suggests that such designed nanomaterials would possibly meet the bandgap requirement for the photocatalytic oxidation of water without an applied electric bias.

8.3.3 Co Nanocrystals and Co_3O_4 Catalysts

The challenges in making isolated cobalt nanocrystals are to overcome the large attractive forces between the nanoparticles due to surface tension and van der Waals interactions that tend to aggregate them.[794,795] The electronic structure of cobalt

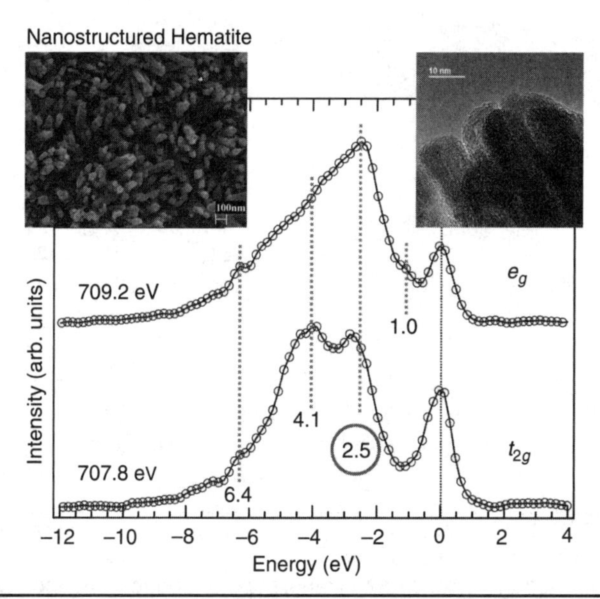

FIGURE 8.6 Energy-dependent resonant inelastic soft x-ray scattering spectra of α-Fe_2O_3 nanorod-arrays. Upper two insets are the electron microscopy images of α-Fe_2O_3 arrays consisting of oriented and bundled ultrafine nanorods.

nanocrystals suspended in liquid as a function of size has been investigated using in situ soft x-ray absorption and emission spectroscopy. A sharp absorption peak associated with the ligand molecules is found that increases in intensity upon reducing the nanocrystal size. X-ray Raman features due to *d-d* and charge-transfer excitations of ligand molecules are identified. The study reveals the local symmetry of the surface of ε-Co phase nanocrystals, which originates from a dynamic interaction between cobalt nanocrystals and surfactant + solvent molecules.[796] Figure 8.7 shows x-ray Raman features in Co L_3-edge RIXS for cobalt nanocrystals of different diameters, and an illustration of the electron transfer from cobalt nanocrystals to the ligand molecules.

Figure 8.7a shows RIXS spectra as a function of Co nanocrystal size at the excitation corresponding to $2p_{3/2}$ absorption edge (778 eV). The −2 eV peak (*d-d* excitation) increases its intensity with nanocrystal size. The unchanged peak position suggests that the local symmetry and surrounding environment in Co nanocrystals remain unchanged. The

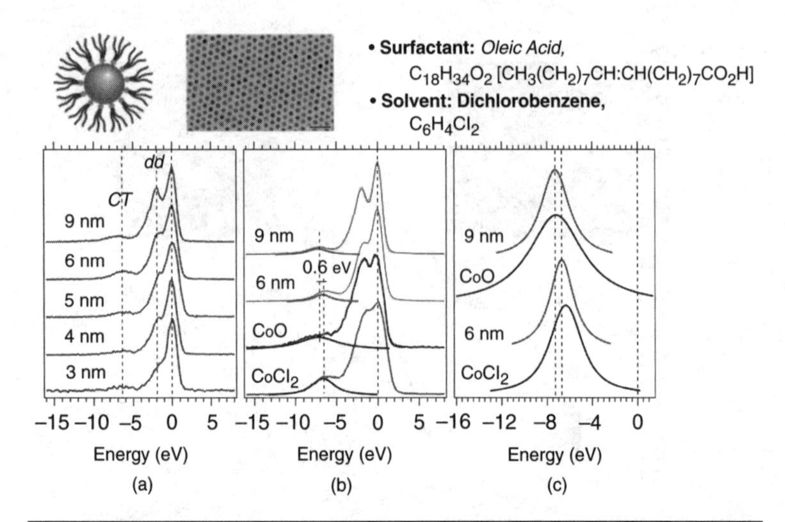

Figure 8.7 RIXS spectra of cobalt nanocrystal suspensions (a); peak fittings of RIXS spectra of cobalt nanocrystal and reference cobalt compounds (b); charge transfer peak of cobalt nanocrystal and cobalt reference compounds (c).

increase of intensity with nanocrystal size is in line with the decreasing fraction of surface Co atoms.

The intensity of the charge-transfer peak centered around –6.7 eV increases with nanocrystal size from 3 to 6 nm and decreases thereafter. Also there is also a peak shift from the 9 to 6 nm nanocrystals. The shift can be accurately determined by deconvolution of the spectra using four Voigt functions corresponding to elastic, *d-d* loss peaks, and metal-ligand charge transfer (MLCT) peaks, as shown in Figure 8.7b. The total fit is excellent and indistinguishable from the experimental data. The charge-transfer peak was found to be at –7.3 eV for the 9 nm Co nanocrystals and at –6.7 eV for 6 nm and smaller ones. The reference spectra of CoO and $CoCl_2$ in Figure 8.7b provide a clue as to the cause to the shift. The CT peak in the 9 nm Co nanocrystals at –7.3 eV is similar to that of CoO, indicating a strong interaction between Co nanocrystals and the carboxyl group of the oleic acid surfactant. For the smaller Co nanocrystals (3–6 nm), the CT peak is found at –6.7 eV, that is in line with $CoCl_2$, which could be the result of penetration of chlorobenzene molecules through the surfactant shell in the smaller nanocrystals due to less efficient packing. The width of CT peaks in the Co nanocrystals is also narrower than that of the CoO and close to that of $CoCl_2$, supporting again that the Co nanocrystals interact with the surrounding $C_6H_4Cl_2$ solvent molecules.

Knowledge of the chemical state of the metal nanoparticles and of the species adsorbed during the Fischer-Tropsch (FT) reaction[797,798] is crucial for a complete understanding of the catalytic process. This is due to a lack of suitable characterization techniques that are able to characterize the chemical and electronic structure of the catalyst during reaction. The in situ soft x-ray absorption study (see in Figure 8.8) shows that the influence of particle size in the carbon monoxide hydrogenation reaction has been studied using cobalt nanoparticles with a narrow size distribution prepared from colloidal chemistry.[799] Model catalysts consisting of cobalt nanoparticles supported on a gold foil were prepared using the Langmuir-Blodgett technique and studied in situ (i.e., interaction conditions of gas pressure and temperature) in the XAS cell.

The spectra at room temperature correspond to CoO, which contains Co^{2+} ions octahedrally coordinated with oxygen

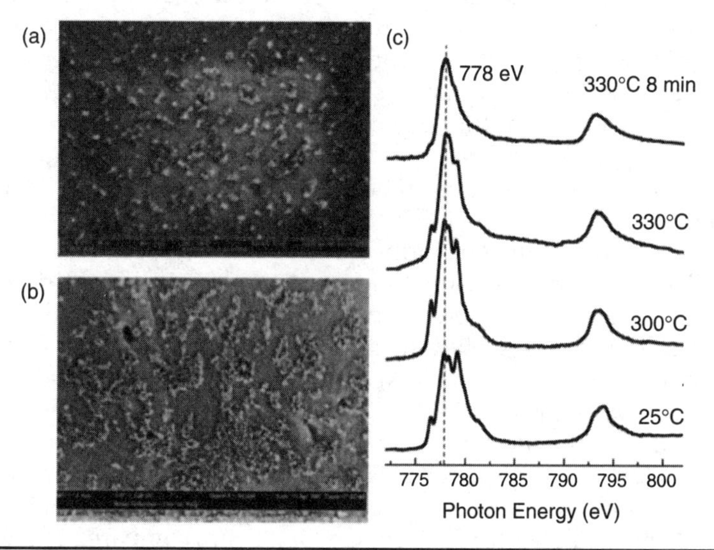

Figure 8.8 High-resolution TEM of (a) cobalt as-prepared supported nanoparticles (10 nm); (b) after reduced process; (c) $L_{2,3}$ x-ray absorption spectra of as-prepared supported nanoparticles in a flow of H_2 (20 mL/min, atmospheric pressure), showing the reduction of samples from CoO to metallic cobalt at elevated temperatures.

anions. The presence of this cobalt oxide is evidenced by the extra features at 776.5 and 780 eV. During the heating treatment in hydrogen from 250 to 330°C, the features in the x-ray absorption spectra changed to those of the metallic state characterized by two asymmetric absorption white lines at 778 and 794 eV. The surfactant covering the NPs after synthesis could be removed by heating to 200 to 270°C in H_2. Soft x-ray absorption spectroscopy was performed using a gas flow cell under reaction conditions of H_2 and CO at atmospheric pressure. 20 Torr of pure hydrogen flow at 330°C removed the protecting surfactant layer and reduced the NPs from oxidized to metallic. The NPs remained metallic during the methanation reaction with their surface covered by CO. The methanation turnover frequency of silica-supported NPs was found to decrease with diameter below 10 nm, whereas the reaction activation energy was found to be independent of NP size.

The challenge in photocatalytic water splitting is finding catalysts that can orchestrate this dance of electrons and protons that follows the creation of electrons and holes by absorption of sunlight. The anode, which links oxygen atoms into oxygen molecules, has been a particularly difficult challenge. Platinum works, but is too expensive and rare to be viable on an industrial scale. Figure 8.9 shows a resonant inelastic soft x-ray scattering study of nanostructured Co_3O_4 in mesoporous silica (SBA15) for water splitting to produce oxygen gas. A charge transfer excitation is identified, and the CT shows a variation in intensity when nanoporous silica is presented.

The Co_3O_4 clusters formed inside the mesoporous silica (SBA-15) are 35 and 70 nm long for 4% and 8% weight loading, respectively. The spheroid shape clusters consist of parallel bundles of nanorods whose structure is imposed by the silica channels. The rods are linked by short bridges, formed by cobalt oxide growth in the micropores interconnecting the mesoscale channels. It is demonstrated that cobalt oxide nanoclusters are good candidates for water oxidation catalysts under mild pH and temperature conditions. O_2 yield is 1,600 times higher for SBA-15/Co_3O_4 (35 nm) compared to that of bare Co_3O_4 micron-sized particles per weight.[800]

The challenge is the constraint imposed by the interplay between the optical, electronic, and chemical properties of the light-absorbing materials. Materials that are stable in

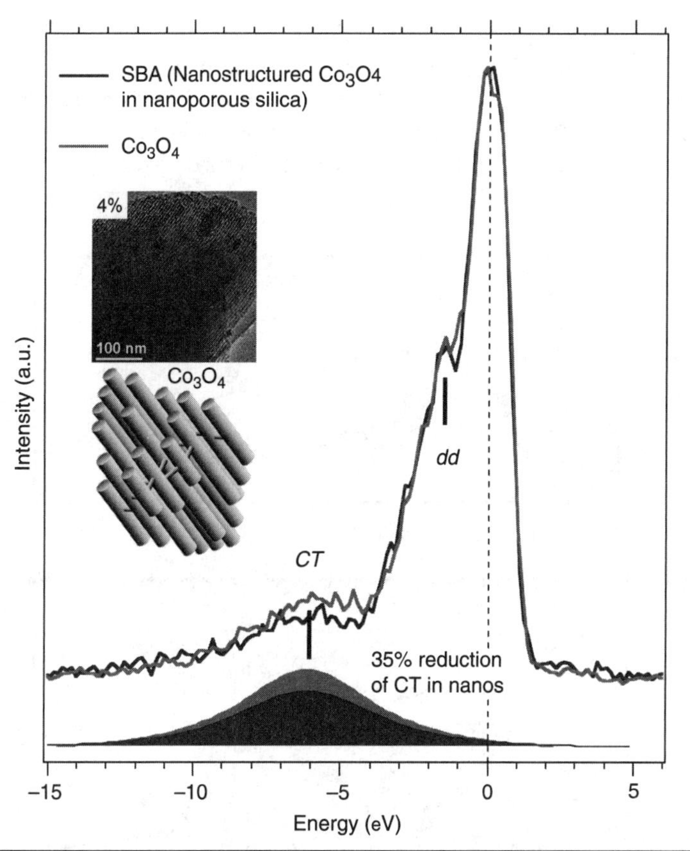

FIGURE 8.9 RIXS spectra of Co_3O_4 microcrystalline powder and nanostructured Co_3O_4 in nanoporous silica (SBA15).

water and that can split water into H_2 and O_2 do not absorb sunlight effectively, and the materials that absorb sunlight effectively cannot sustain photochemically induced water splitting. Soft x-rays are ideal for studying electron transitions between a localized core state and a valence/conduction state; thus, has unique features that make it a powerful tool to extract information about electronic properties. Photon-in/photon-out soft x-ray spectroscopy can be used to study different chemical species of interest in depth under the real-world conditions, such as the presence of electric and/or magnetic fields and/or under gas catalytic reaction/or liquid and electrochemical reaction conditions.

8.3.4 ZnO

ZnO has nearly the same bandgap and electron affinity as TiO_2, making it a possible candidate as an effective dye sensitized solar cell (DSSC) semiconductor. While little work has been done on large-scale, template-free growth of TiO_2 nanowires,

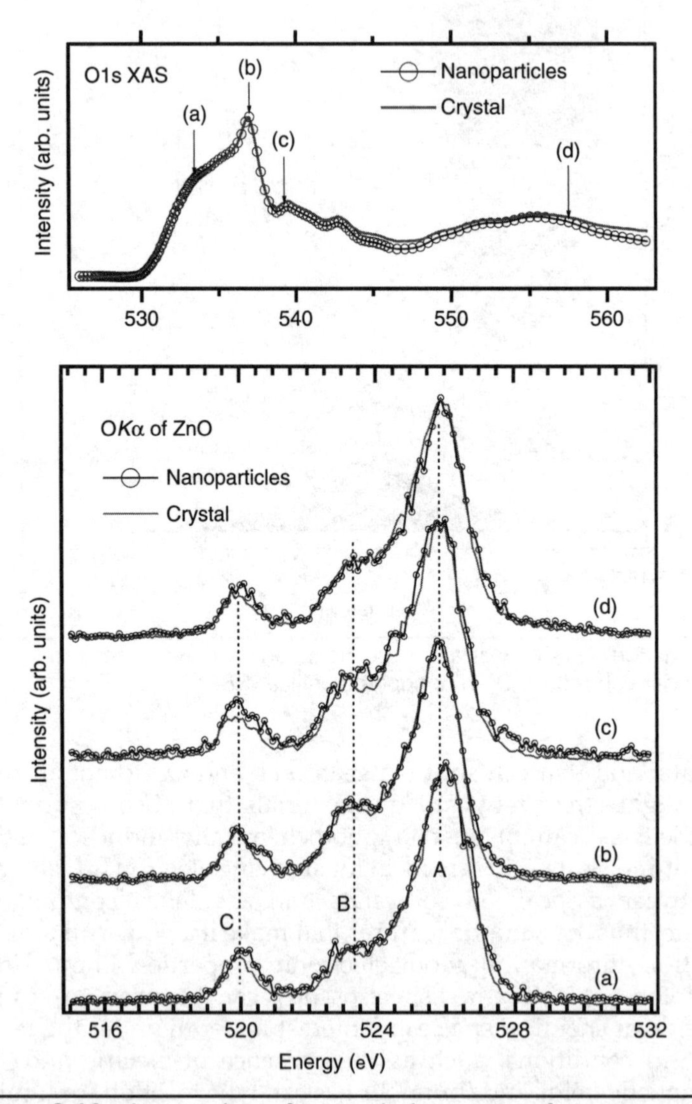

FIGURE 8.10 A comparison of x-ray emission spectra of nanostructured and bulk crystalline ZnO recorded at selected excitation energies of (a), (b), (c), and (d) which are indicated in the XAS spectra (upper panel).

ZnO can readily be grown in a variety of morphologies and by several different processing methods. There are already many soft x-ray spectroscopic experiments performed on ZnO crystals, nanocrystals, and films, and also doped ZnO.[801,802]

The XES spectra of bulk and nanostructured ZnO are displayed together with the corresponding XAS spectrum in Figure 8.10. The O K emission spectra reflect the O 2*p* occupied states (valence band), and the O 1*s* absorption spectrum reflects the O 2*p* unoccupied states (conduction band). In the photon energy region of 530 to 539 eV, the x-ray absorption can be mainly assigned to the O 2*p* hybridized with Zn 4*s* states. In the region of 539 to 550 eV the spectrum is mainly attributed to O 2*p* hybridized with Zn 4*p* states. Above 550 eV, the contribution is mainly coming from O 2*p*–Zn 4*d* mixed states.[801] Stronger $s - p - d$ hybridization was revealed in nanostructured ZnO since the contributions of features at 520 eV and 523 eV are enhanced. A well-defined bandgap can be observed between the valence-band maximum and conduction-band minimum. The *absorption–emission* spectrum yields the fundamental bandgap energy of 3.3 eV, which is in a good agreement with the 3.4 eV found for bulk ZnO.[803]

The O K emission spectra of ZnO bulk and nanoparticles, recorded at different excitation energies, are displayed in Figure 8.10. The selected excitation energies are indicated in the XAS spectrum. By selecting different excitation energies, predominant contribution of specific admixture of unoccupied O 2*p* with Zn 3*d*, 4*s*, and 4*p* states is expected. Indeed, three distinct structures can be observed in the XES spectra, labeled as A, B, and C. Feature A, located at 526 eV, is mainly due to O 2*p*–Zn 4*p* states. The low energy shoulder (feature B) in 522–524 eV region arises from the mixed states of O 2*p*–Zn 4*s*. The single-band shape at 520 eV is attributed mainly to the O 2*p* hybridized with Zn 3*d* states.

8.4 In Situ Electronic Structure Characterization

Performing in situ studies with these techniques will enable us to understand the unique catalytic property associated with these ultrafine catalysts. An in situ characterization of catalytic reactions of sub-nanometer Pt NPs is illustrated in Figure 8.11. There are two directions to pursue in the study:

1. Understand the oxidation tendency of metal atoms in NPs with size below 1 nm. It was observed that the platinum atoms are more oxidized than the rhodium

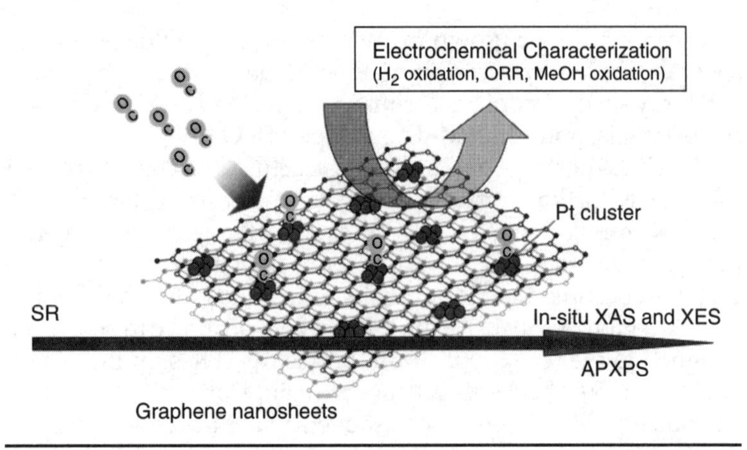

FIGURE 8.11 Characterization of H_2, CO, and/or MeOH oxidations.

atoms in the 1 nm Pt/Rh bimetallic NPs, which is contrary to the common belief that rhodium has a higher oxygen affinity than platinum. The question is whether this reversed oxidation tendency is general for the bimetallic NPs with size below 1 nm. To answer this question, the bimetallic NPs of metals with large disparity in oxygen affinity, such as Pt/Pd, Pt/Ru, and Pt/Co, will be studied. The oxidation states of the two metal components can be monitored by photon-in/photon-out soft x-ray spectroscopy during in situ reduction and oxidation.

2. Establish the correlation between catalytic activity and the surface structure of NPs. Two benchmark catalytic reactions, CO oxidation and ethylene hydrogenation over bimetallic NPs, need to be studied in an in situ reactor that allows performing the turnover measurement and the photon-in/photon-out soft x-ray spectroscopy. This study will help to understand how the oxidation states of metals and the surface-alloying structure play their roles in the catalytic processes over these ultrafine bimetallic NPs. Reactions with multiple products such as pyrrole hydrogenation will be studied in order to explore the possibility of controlling reaction selectivity by changing the surface-alloying structure of NPs.

8.5 $NiCl_2$ Water Solutions

Most properties of electrolyte solutions depend on the ability of solvent and solute to interact, which is the nature of the complex ion formation. One important parameter is the Gibbs free energy of solvation, which requires an assumption of the effective ionic radii ${}^{eff}R_{ion}$ that is often expressed by $R_{ion} + \Delta R$. Here, ΔR is a function taking into account the first hydration shell.[804] Due to the importance of ion behavior in electrolyte solutions, ions in electrolytes have been commonly studied in many theoretical and experimental investigations. For Ni^{2+} in aqueous solution, neutron diffraction studies[805,806] as well as x-ray diffraction[807] have shown that Ni^{2+} has a coordination sphere of six water molecules. Inner sphere contact pairs of Ni-Cl have been suggested to exist for 8 percent of the Ni ions and ion-pair formation in x-ray diffraction study of $NiCl_2$ and $NiBr_2$ aqueous solutions.[808,809]

Figure 8.12a shows the XAS study of aqueous $NiCl_2$ solutions as a function of electrolyte concentrations from 50 to 1500 mM; a systematic change in the XAS spectral features is observed. One would expect a transition in the importance of inter ionic interactions. At the L_3 edge, two peaks, P1 and P2, split by 1.6 eV are expected in solid-state $NiCl_2$. The satellite peak (P2) is known to be due to the ligand metal charge transfer

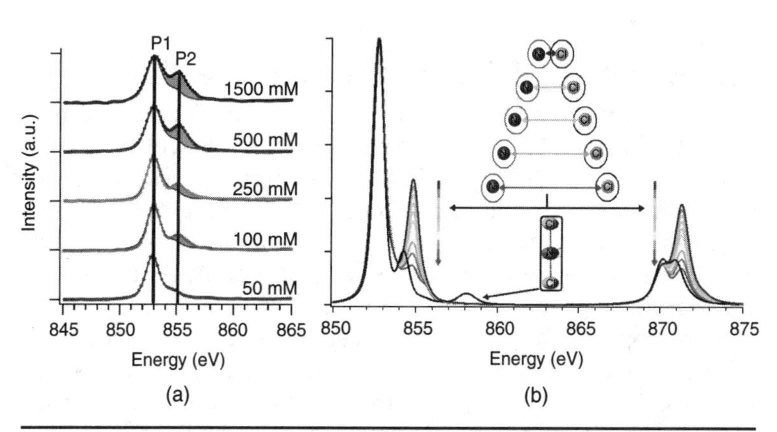

FIGURE 8.12 Nickel L_3-edge FY-detection XAS spectra for $NiCl_2$ water solutions as a function of the concentrations (a); electronic structure calculations of Ni(II) atoms in the solution changes as a function of Ni-Cl distances (b).

(LMCT) from Cl^- to Ni^{2+},[810–811] which is made possible by the close proximity of Ni and Cl in the solid.

The energy splitting between the first two absorption peaks within the L_3 edge (P1 and P2) is increased from 1.6 eV in the solid to 2.4 eV in the electrolytes. With increasing electrolyte concentration, the P2/P1 intensity ratio increases. The changes in peak energies and intensities are reproduced by a multiplet calculation of the Ni XAS spectra in O_h symmetry (with a d^8 electron configuration) by a variation of the relative contributions of the dipole matrix elements with different symmetry. This is equivalent to an increased contribution of singlet states versus the triplet states for increasing electrolyte concentration. These changes in the electronic structure are related to the increasing importance of solvent-shared ion pairs at elevated electrolyte concentrations, manifesting itself in a progressive distortion of the local O_h symmetry around the Ni ions.

Clearly, this geometrical saturation effect alone cannot account for the observed intensity changes as a function of electrolyte concentration. It is the remaining effect of concentration, which is evident from the change of XAS spectral shape, that the electronic structure locally at the Ni(II) atoms in the solution changes as a function of electrolyte concentration. These changes in the electronic structure are rationalized on the basis of electronic structure calculations (see Figure 8.12b).

8.6 In Situ Study of Electrochemical Reaction

Understanding the mechanism of electrochemical reactions is important from a basic science point of view, as well as for the development of electrochemical applications, such as batteries, solar cells, and fuel cells, and so on.[19, 57, 812–815] It is also important in applications of the corrosion and wear of copper brushes in electrical motors.[816] This is best accomplished by the combination of electrochemical measurements with x-ray spectroscopic techniques. However, ex situ characterization methods suffer from a number of experimental limitations. For instance, transferring samples from one instrument to another can easily result in chemical changes in the state of the sample and introduce artifacts. In order to avoid the unwanted change of composition and electronic structure, in situ techniques are required. Several in situ techniques have been used to investigate electrochemical processes, including scanning probe microscopy, x-ray diffraction, infrared reflection absorption spectroscopy, surface-enhanced Raman spectroscopy, sum frequency generation, and so on.[817,818] However, these techniques have limited capabilities in providing direct electronic structure information, which is the key for monitoring changes in the oxidation state.

Figure 8.13 shows the study of the electronic structure and formation dynamics of the two different copper oxides (CuO, cupric oxide and Cu_2O, cuprous oxide) as a function of the electrode potential.[819] This setup allows the real-time monitoring of the element-specific electronic structure changes of an interfacial system at all stages of the electrochemical cycle. Using this technique we have investigated the oxidation of Cu in $NaHCO_3$ solution under controlled electrochemical conditions.

The cyclic voltammetric curve of the Cu thin film shows a scan rate of 0.02 V/s in 0.1 M $NaHCO_3$ solution in this cell setup. The characteristic oxidation and reduction peaks of copper are clearly observed. Anodic peaks I and II correspond to the formation of Cu^+ and Cu^{2+}, respectively. Cathodic peaks II and I are related to the reduction of Cu^{2+} to Cu^+ and Cu^+ to Cu^0, respectively. These observations and assignments

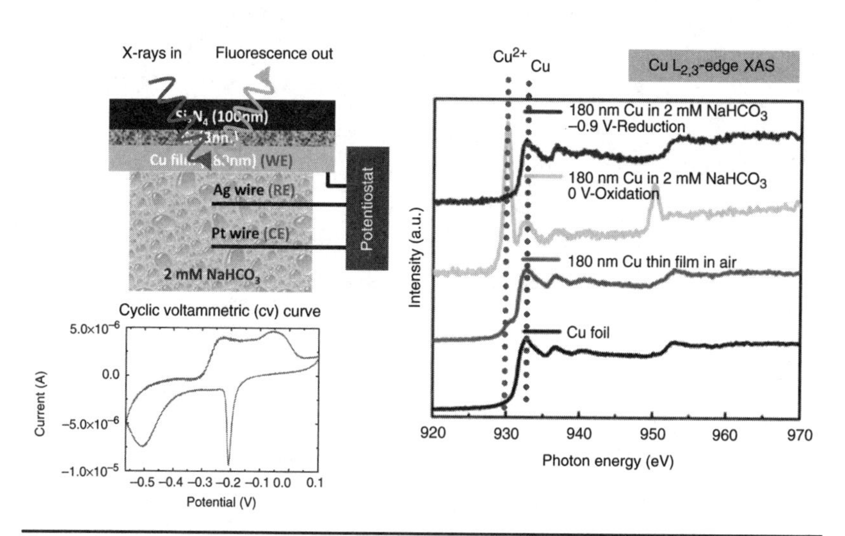

FIGURE 8.13 Left upper: illustration of electrochemical cell; Left bottom: Cu redox curve in 2 mM Na_2HCO_3 water solution; Right: XAS spectra of Cu electrode under oxidation/reduction condition.

are consistent with previous reports.[820,821] Cu $L_{2,3}$-edge XAS of Cu thin film on Si_3N_4 membrane before contacting any solution is shown in Figure 8.13b. The main component can be assigned to the L_3-edge (932.7 eV) of the metallic Cu, consistent with the metallic Cu reference sample (Figure 8.13 curve (1)).[822–824] There is one additional peak other than the main peak at the lower photon energy side (930 eV) that can be ascribed to the Cu^{2+} species. These features indicate that a thin oxidized layer is formed on the surface of the evaporated Cu film due to oxidation in air during transfer. These observations indicate the importance of in situ measurements for investigations of electrochemical reactions in order to avoid introducing artifacts and misinterpreting experimental data.

References

1. G. W. Crabtree and N. S. Lewis, Solar Energy Conversion, *Physics Today*, March 2007.
2. J. R. Bolton, S. J. Strickler, and J. S. Connolly, *Nature* **316**, 495 (1985).
3. A. J. Bard, Photoelectrochemistry, *Science* **207**, 139 (1980).
4. J. R. Bolton, Solar Fuels, *Science* **202**, 705 (1978).
5. G. Ciamician, *Science* **36**, 642 (1912).
6. Arthur J. Esswein and Daniel G. Nocera, *Chem. Rev.* **107**, 4022–4047 (2007).
7. A. Fujishima and K. Honda, *Nature* **238**, 37 (1972).
8. T. Inoue, A. Fujishima, S. Konishi, and K. Honda, *Nature* **277**, 637 (1979).
9. Z. Zhou, J. Ye, K. Sayama, and H. Arakawa, *Nature* **414**, 625 (2001).
10. A. Paets van Troostwijk and J. R. Deiman, *Obs. Phys.* **35**, 369 (1789).
11. A. Paets van Troostwijk and J. R. Deiman, *J. Phys.* **2**, 130 (1790).
12. A. Paets van Troostwijk and Algem. Mag. Wetensch. *Kunst Smaak* **4**, 909 (1790).
13. F. A. C. Gren, *J. Phys.* **2**, 194 (1790).
14. G. Pearson, *Phil. Trans.* **87**, 142 (1797).
15. R. de Levie, *Journal of Electroanalytical Chemistry* **476**, 9293 (1999).
16. W. Nicholson, Nicholson's *J. Nat. Phil. Chem. Arts* **4**, 179 (1800).
17. W. Nicholson, *Ann. Phys.* **6**, 340 (1800).
18. Linus Pauling, *General Chemistry*, Section 15–2. San Francisco, 1970.
19. M. W. Kanan, and D. G. Nocera, *Science* **321**, 1072 (2008).

20. Werner Zittel and Reinhold Wurster (1996). "Chapter 3: Production of Hydrogen. Part 4: Production from electricity by means of electrolysis." *HyWeb: Knowledge—Hydrogen in the Energy Sector*. Ludwig-Bölkow-Systemtechnik GmbH.
21. A. Kudo and Y. Miseki, *Chem. Soc. Rev*. **38**, 253–278 (2009).
22. A. Fujishima, T. N. Rao, and D. A. Tryk, *J. Photochem. Photobiol.* C **1**, 1 (2000).
23. A. Fujishima, X. Zhang, and D. A. Tryk, *Int. J. Hydrogen Energy* **32**, 2664 (2007).
24. M. Grätzel, *Energy Resources through Photochemistry and Catalysis*, Academic Press, New York, 1983.
25. N. Serpone and E. Pelizzetti, *Photocatalysis*, Wiley, New York, 1989.
26. H. Yoneyama, *Crit. Rev. Solid State Mater. Sci*. **18**, 69 (1993).
27. K. Domen, J. N. Kondo, M. Hara, and T. Takata, *Bull. Chem. Soc. Jpn*. **73**, 1307 (2000).
28. H. Arakawa and K. Sayama, *Catal. Surv. Jpn*. **4**, 75 (2000).
29. K. Domen, M. Hara, J. N. Kondo, T. Takata, A. Kudo, H. Kobayashi, and Y. Inoue, *Korean J. Chem. Eng*. **18**, 862 (2001).
30. A. Kudo, *J. Ceram. Soc. Jpn*. **109**, 81 (2001).
31. A. Kudo, *Catal. Surv. Asia*, **7**, 31 (2003).
32. Z. Zou and H. Arakawa, *J. Photochem. Photobiol.* A **158**, 145 (2003).
33. H. Yamashita, M. Takeuchi, and M. Anpo, *Encyclopedia of Nanoscience and Nanotechnology*, American Scientific Publishers, California, 2004, 10, p. 639.
34. M. Anpo, S. Dohshi, M. Kitano, Y. Hu, M. Takeuchi, and M. Matsuoka, *Annu. Rev. Mater. Res*. **35**, 1 (2005).
35. J. S. Lee, *Catal. Surv. Asia* **9**, 217 (2005).
36. A. Kudo, *Int. J. Hydrogen Energy* **31**, 197 (2006).
37. Y. Inoue, *Chem. Ind*. **108**, 623 (2006).
38. W. Zhang, S. B. Park, and E. Kim, *Energy and Fuel* 295, 2006.
39. K. Maeda, K. Teramura, N. Saito, Y. Inoue, H. Kobayashi, and K. Domen, *Pure Appl. Chem*. **78**, 2267 (2006).
40. W. Shanguan, *Sci. Technol. Adv. Mater*. **8**, 76 (2007).
41. A. Kudo, *Pure Appl. Chem*. **79**, 1917 (2007).
42. A. Kudo, *Int. J. Hydrogen Energy* **32**, 2673 (2007).
43. K. Maeda and K. Domen, *J. Phys. Chem.* C **111**, 7851 (2007).
44. K. Maeda, K. Teramura, and K. Domen, *Catal. Surv. Asia* **11**, 145 (2007).
45. S. Ekambaram, *J. Alloys Compd*. **448**, 238 (2008).
46. M. Laniecki, *Ceram. Eng. Sci. Proc*. **28**, 23 (2008).

47. F. E. Osterloh, *Chem. Mater.* **20**, 35 (2008).
48. J. Nozik, *Annu. Rev. Phys. Chem.* **29**, 189 (1978).
49. Y. V. Pleskov and Y. Y. Gurevich, in *Semiconductor Photoelectrochemistry*, ed. P. N. Bartlett, Plenum, New York, 1986.
50. N. S. Lewis and D. G. Nocera, *PNSA* **103**, 15729 (2006).
51. A. J. Bard and M. A. Fox, *Acc. Chem. Res.* **28**, 141 (1995).
52. T. A. Betley, Q. Wu, and T. Van Voorhis, *D. G. Nocera, Inorg. Chem.* **47**, 1849 (2008).
53. R. I. Cukier and D. G. Nocera, *Annu. Rev. Phys. Chem.* **49**, 337 (1998).
54. M. H. V. Huynh and T. J. Meyer, *Chem. Rev.* **107**, 5004 (2007).
55. R. Eisenberg and H. B. Gray, *Inorg. Chem.* **47**, 1697 (2008).
56. Junko Yano, Yulia Pushkar, Pieter Glatzel, Azul Lewis, Kenneth Sauer, Johannes Messinger, Uwe Bergmann, and Vittal Yachandra, *J Am. Chem. Soc.* **127**, 14974 (2005).
57. Junko Yano, Jan Kern, Kenneth Sauer, Matthew J. Latimer, Yulia Pushkar, Jacek Biesiadka, Bernhard Loll, Wolfram Saenger, Johannes Messinger, Athina Zouni, and Vittal K. Yachandra, *Science* **314**, 821 (2006).
58. T. Inoue, A. Fujishima, S. Konishi, and K. Honda, *Nature* **277**, 637 (1979).
59. Z. Zhou, J. Ye, K. Sayama, and H. Arakawa, *Nature* **414**, 625 (2001).
60. J. C. Ireland, P. Klostermann, E. W. Rice, and R. M. Clark, *Appl. Environ. Microbiol.* **59**, 1668–1670 (1993).
61. J. C. Sjogren and R. A. Sierka, *Appl. Environ. Microbiol.* **60**, 344–347 (1994).
62. R. X. Cai, Y. Kubota, T. Shuin, H. Sakai, K. Hashimoto, and A. Fujishima, *Cancer Res.* **52**, 2346–2348 (1992).
63. R. Cai, K. Hashimoto, Y. Kubota, and A. Fujishima, *Chem. Lett.* 427–430 (1992).
64. K. Kalyanasundaram, E. Borgarello, D. Duonghong, M. Gratzel, Angew. *Chem., Int. Ed. Engl.* **20**, 987 (1981).
65. D. Duonghong, E. Borgarello, and M. Gratzel, *M. J. Am. Chem. Soc.* **103**, 4685 (1981).
66. E. Borgarello, J. Kiwi, E. Pelizzetti, M. Visca, and M. Gratzel, *Nature* **289**, 158 (1981).
67. A. Wold, *Chem. Mater.* **5**, 280–283 (1993).
68. K. Suzuki, In *Photocatalytic Purification and Treatment of Water and Air*, D. F. Ollis, H. Al-Ekabi, Eds., Elsevier: Amsterdam, 1993.

69. K. E. Karakitsou and X. E. Verykios, *J. Phys. Chem.* **97**, 1184–1189 (1993).
70. M. Gratzel, *ACCC. hem. Res.* **14**, 376 (1981).
71. M. Khan and N. N. Rao, *J. Photochem. Photobiol. A: Chem.* **56**, 101–111 (1991).
72. M. Khan, D. Chatterjee, M. Krishnaratnam, and M. Bala, *J. Mol. Catal.* **72**, 13–18 (1992).
73. M. Khan, D. Chatterjee, and M. Bala, *J. Photochem. Photobiol. A: Chem.* **67**, 349–352 (1992).
74. H. Gerischer and A. Heller, *J. Electrochem. Soc.* **139**, 113–118 (1992).
75. N. B. Jackson, C. M. Wang, Z. Luo, J. Schwitzgebel, J. G. Ekerdt, J. R. Brock, and A. J. Heller, *Electrochem. Soc.* **138**, 3660–3664 (1991).
76. M. Nair, Z. H. Luo, and A. Heller, *Ind. Eng. Chem. Res.* **32**, 2318–2323 (1993).
77. K. W. Boer, *Survey of Semiconductor Physics*, Van Nostrand Reinhold, New York, 1990, p.249.
78. G. Rothenberger, J. Moser, M. Gratzel, N. Serpone, and D. K. Sharma, *J. Am. Chem. Soc.* **107**, 8054 (1985).
79. M. Grätzel, *Heterogeneous Photochemical Electron Transfer*, CRC Press: Boca Raton, FL, 1989.
80. R. Memming, In *Topics in Current Chemistry*; Steckham, E., Ed.; Springer-Verlag: Berlin, 1988; Vol. 143, pp 79–113.
81. E. R. Carraway, A. J. Hoffman, and M. R. Hoffmann, *Environ. Sci. Technol.* **28**, 786–793 (1994).
82. C. Kormann, D. W. Bahnemann, and M. R. Hoffmann, *Environ. Sci. Technol.* **22**, 798–806 (1988).
83. A. J. Hoffmann, E. R. Carraway, and M. R. Hoffmann, *Environ. Sci. Technol.* **28**, 776–785 (1994).
84. C. M. Wang, A. Heller, and H. Gerischer, *J. Am. Chem. Soc.* **114**, 5230–5234 (1992).
85. H. Gerischer and A. Heller, *J. Phys. Chem.* **95**, 5261–5267 (1991).
86. S. Pehkonen, R. Siefert, S. Webb, and M. R. Hoffmann, *Environ. Sei. Technol.* **26**, 2056 (1993).
87. M. V. Rao, K. Rajeshwar, V. Vernerker, and J. Dubow, *J. Phys. Chem.* **84**, 1987 (1980).
88. S. Nishimoto, B. Ohtani, H. Kajiwara, and T. Kagiya, *J. Chem. Soc., Faraday Trans.* **81**, 61 (1985).
89. D. W. Bahnemann, C. Kormann, and M. R. Hoffmann, *J. Phys. Chem.* **91**, 3789–3798 (1987).

90. S. T. Martin, H. Herrmann, W. Choi, and M. R. Hoffmann, *Trans. Faraday Soc.* **90**, 3315–3323 (1994).
91. S. T. Martin, H. Herrmann, and M. R. Hoffmann, *Trans. Faraday Soc.* **90**, 3323–3330 (1994).
92. A. Mills, R. H. Davies, and D. Worsley, *Chem. Soc. Rev.* **22**, 417–425 (1993).
93. A. Mills, S. Morns, and R. Davies, *J. Photochem. Photobiol. A: Chem.* **70**, 183 (1993).
94. S. T. Martin, C. L. Morrison, and M. R. Hoffmann, *J. Phys. Chem.* **98**, 13695 (1994).
95. H. Noda, K. Oikawa, and H. Kamada, *Bull. Chem. Soc. Jpn.* **66**, 455–458 (1993).
96. B. Ohtani and S. Nishimoto, *J. Phys. Chem.* **97**, 920–926 (1993).
97. B. V. Mihaylov, J. L. Hendrix, and J. H. Nelson, *J. Photochem. Photobiol. A: Chem.* **72**, 173–177 (1993).
98. X. Domenech, In *Photocatalytic Purification and Treatment of Water and Air*; D. F. Ollis, H. Al-Ekabi, Eds.; Elsevier: Amsterdam, 1993; pp 337–351.
99. J. Peral, J. Munoz, and J. Domenech, *Photochem. Photobiol.* **55**, 251 (1990).
100. J. Peral and X. Domenech, *J. Chem. Technol. Biotechnol.* **53**, 93–96 (1992).
101. K. Tanaka, T. Hisanaga, and A. P. Rivera, In *Photocatalytic Purification and Treatment of Water and Air*; H. Al-Ekabi, D. F. Ollis, Eds.; Elsevier: Amsterdam, 1993; pp 169–178.
102. S. Sato and J. M. White, *Chem. Phys. Lett.* **72**, 83 (1980).
103. J.-M. Lehn, J.-P. Sauvage, and R. Ziessel, *Nouv. J. Chim.* **4**, 623 (1980).
104. K. Yamaguti and S. Sato, *J. Chem. Soc., Faraday Trans.* 1 **81**, 1237 (1985).
105. G. R. Bamwenda, S. Tshbota, T. Nakamura, and M. Haruta, *J. Photochem. Photobiol.* A **89**, 177 (1995).
106. A. Iwase, H. Kato, and A. Kudo, *Catal. Lett.* **108**, 7 (2006).
107. K. Domen, S. Naito, S. Soma, M. Onishi, and K. Tamaru, *J. Chem. Soc. Chem. Commun.* 543 (1980).
108. T. Kawai and T. Sakata, *Chem. Phys. Lett.* **72**, 87 (1980).
109. Y. Inoue, O. Hayashi, and K. Sato, *J. Chem. Soc. Faraday Trans.* **86**, 2277 (1990).
110. K. Maeda, K. Teramura, D. Lu, N. Saito, Y. Inoue, and K. Domen, *Angew. Chem. Int. Ed.* **45**, 7806 (2006).
111. K. Maeda, K. Teramura, D. Lu, N. Saito, Y. Inoue, and K. Domen, *J. Catal.* **243**, 303 (2006).

112. A. Iwase, H. Kato, and A. Kudo, *Chem. Lett.* **53**34, 946 (2005).
113. M. Hara, C. C. Waraksa, J. T. Lean, B. A. Lewis and T. E. Mallouk, *J. Phys. Chem.* A **104**, 5275 (2000).
114. A. Ishikawa, T. Takata, J. N. Kondo, M. Hara, H. Kobayashi, and K. Domen, *J. Am. Chem.* Soc. **124**, 13547 (2002).
115. G. N. Schrauzer and T. D. Guth, *J. Am. Chem. Soc.* **99**, 7189 (1977).
116. K. Sayama and H. Arakawa, *J. Chem. Soc., Faraday Trans.* **93**, 1647 (1997).
117. J. Shi, J. Chen, Z. Feng, T. Chen, Y. Lian, X. Wang, and C. Li, *J. Phys. Chem. C* **111**, 693 (2007).
118. J. Zhang, Q. Xu, Z. Feng, M. Li, and C. Li, *Angew. Chem. Int. Ed.* **133**47, 1766 (2008).
119. A. Kudo, K. Domen, K. Maruya, and T. Onishi, *Chem. Phys. Lett.* **133**, 517 (1987).
120. S. Tabata, H. Nishida, Y. Masaki, and K. Tabata, *Catal. Lett.* **34**, 245 (1995).
121. X. Chen, *Chin. J. Catal.* **30**, 839 (2009).
122. A. Kudo, K. Domen, K. Maruya, and T. Onishi, *Chem. Lett.* **16**, 1019 (1987).
123. K. Sayama and H. Arakawa, *J. Photochem. Photobiol. A: Chem.* **77**, 243 (1994).
124. J. Chae, J. Lee, J. H. Jeong, and M. Kang, *Bull. Korean Chem. Soc.* **30**, 302 (2009).
125. D. Jing, Y. Zhang, L. Guo, *Chem. Phys. Lett.* **415**, 74 (2005).
126. R. Sasikala, V. Sudarsan, C. Sudakar, R. Naik, T. Sakuntala, S. R. Bharadwaj, *Int. J. Hydrogen Energy* **33**, 4966 (2008).
127. M. Zalas, M. La, *Sol. Energy Mater. Sol. Cells* **89**, 287 (2005).
128. R. Sasikala, A. Shirole, V. Sudarsan, T. Sakuntala, C. Sudakar, R. Naik, S. R. Bharadwaj, *Int. J. Hydrogen Energy* **34**, 3621 (2009).
129. Q. Yuan, Y. Liu, L. L. Li, Z. X. Li, C. J. Fang, W. T. Duan, X. G. Li, C. H. Yan, *Micropor. Mesopor. Mater.* **124**, 169 (2009).
130. S. Xu, D. D. Sun, *Int. J. Hydrogen Energy* **34**, 6096 (2009).
131. S. Xu, J. Ng, X. Zhang, H. Bai, D. D. Sun, *Int. J. Hydrogen Energy* **35**, 5254 (2010).
132. H. J. Choi, M. Kang, *Int. J. Hydrogen Energy* **32**, 3841 (2007).
133. J. W. Park, M. Kang, *Int. J. Hydrogen Energy* **32**, 4840 (2007).
134. K. Lalitha, J. K. Reddy, M. V. P. Sharma, V. D. Kumari, M. Subrahmanyam, *Int. J. Hydrogen Energy* **35**, 3991 (2010).
135. M. S. Park, *M.* Kang, *Mater. Lett.* **62**, 183 (2008).
136. B. Zielinska, E. Borowiak-Palen, R. J. Kalenczuk, *Int. J. Hydrogen Energy* **33**, 1797 (2008).

137. S. C. Moon, H. Mametsuka, E. Suzuki, M. Anpo, *Chem. Lett.* **27**, 117 (1998).
138. S. C. Moon, H. Mametsuka, S. Tabata, E. Suzuki, *Catal. Today* **58**, 125 (2000).
139. S. H. Liu, H. P. Wang, Y. J. Huang, Y. M. Sun, K. S. Lin, M. C. Hsiao, *Energy Source* **25**, 591 (2003).
140. D. Zhao, S. Budhi, A. Rodriguez, R. T. Koodali, *Int. J. Hydrogen Energy* **35**, 5276 (2010).
141. M. Shibata, A. Kudo, A. Tanaka, K. Domen, K. Maruya, T. Ohishi, *Chem. Lett.* **16**, 1017 (1987).
142. M. R. Allen, A. Thibert, E. M. Sabio, N. D. Browning, D. S. Larsen, F. E. Osterloh, *Chem. Mater.* **22**, 1220 (2010).
143. M. Machida, X. W. Ma, H. Taniguchi, J. Yabunaka, T. J. Kijima, *Mol. Catal. A: Chem.* **155**, 131 (2000).
144. J. S. Jang, S. H. Choi, D. H. Kim, J. W. Jang, K. S. Lee, J. S. Lee, *J. Phys. Chem. C* **113**, 8990 (2009).
145. A. Kudo, T. J. Kondo, *Mater. Chem.* **7**, 777 (1997).
146. S. Ogura, M. Kohno, K. Sato, Y. Inoue, *Appl. Surf. Sci.* **121–122**, 521–524 (1997).
147. Y. Inoue, T. Kubokawa, K. Sato, *J. Chem. Soc. Chem. Commun.* **19**, 1298–1299 (1990).
148. Y. Inoue, T. Kubokawa, K. Sato, *J. Phys. Chem.* **95**, 4059 (1991).
149. S. Ogura, K. Sato, Y. Inoue, *Phys. Chem. Chem.* Phys. **2**, 2449 (2000).
150. M. Kohno, T. Kaneko, S. Ogura, K. Sato, Y. Inoue, *J. Chem. Soc. Faraday Trans.* **94**, 89 (1998).
151. Y. Inoue, T. Niiyama, Y. Asai, K. Sato, *J. Chem. Soc. Chem. Commun.* **7**, 579 (1992).
152. Y. Inoue, Y. Asai, K. Sato, *J. Chem. Soc. Faraday Trans.* **90**, 797 (1994).
153. Y. Yamashita, K. Yoshida, M. Kakihana, S. Uchida, T. Sato, *Chem. Mater.* **11**, 61 (1999).
154. K. Domen, S. Naito, T. Onishi, T. Tamaru and M. Soma, *J. Phys. Chem.* **86**, 3657 (1982).
155. K. Domen, S. Naito, T. Onishi and K. Tamaru, *Chem. Phys.* Lett. **92**, 433 (1982).
156. K. Domen, A. Kudo, T. Onishi, *J. Catal.* **102**, 92 (1986).
157. A. Kudo, A. Tanaka, K. Domen, T. Onishi, *J. Catal.* **111**, 296 (1988).
158. K. Domen, A. Kudo, T. Onishi, N. Kosugi, H. Kuroda, *J. Phys. Chem.* **90**, 292 (1986).

159. Y. Liu, L. Xie, Y. Li, R. Yang, J. Qu, Y. Li, X. Li, *J. Power Sources* **183**, 701 (2008).
160. Y. Qin, G. Wang, Y. Wang, *Catal. Commun.* **8**, 926 (2007).
161. T. Takata, K. Domen, *J. Phys. Chem.* C **113**, 19386 (2009).
162. H. Jeong, T. Kim, D. Kim, K. Kim, *Int. J. Hydrogen Energy* **31**, 1142 (2006).
163. Y. G. Ko, W. Y. Lee, *Catal. Lett.* **83**, 157 (2002).
164. H. Mizoguchi, K. Ueda, M. Orita, S. C. Moon, K. Kajihara, M. Hirano, H. Hosono, *Mater. Res. Bull.* **37**, 2401 (2002).
165. W. Sun, S. Zhang, C. Wang, Z. Liu, Z. Mao, *Catal. Lett.* **119**, 148 (2007).
166. A. Kim, D. W. Hwang, S. W. Bae, Y. G. Kim, J. S. Lee, *Korean J. Chem. Eng.* **18**, 941 (2001).
167. J. Kim, D. W. Hwang, H. G. Kim, S. W. Bae, S. M. Ji, J. S. Lee, *Chem. Commun.* **21**, 2488 (2002).
168. H. G. Kim, S. M. Ji, J. S. Jang, S. W. Bae, J. S. Lee, *Korean J. Chem.* Eng. **21**, 970 (2004).
169. J. Kim, D. W. Hwang, H. G. Kim, S. W. Bae, J. S. Lee, W. Li, S. H. Oh, *Top. Catal.* **35**, 295 (2005).
170. Z. Li, G. Chen, X. Tian, Y. Li, *Mater. Res. Bull.* **43**, 1781 (2008).
171. D. W. Hwang, J. S. Lee, W. Li, S. H. Oh, *J. Phys. Chem.* B **107**, 4963 (2003).
172. R. Abe, M. Higashi, Z. Zou, K. Sayama, Y. Abe, *Chem. Lett.* **33**, 954 (2004).
173. M. Higashi, R. Abe, K. Sayama, H. Sugihara, Y. Abe, *Chem. Lett.* **34**, 1122 (2005).
174. R. Abe, M. Higashi, K. Sayama, Y. Abe, H. Sugihara, *J. Phys. Chem. B* **110**, 2219 (2006).
175. M. Uno, A. Kosuga, M. Okui, K. Horisaka, S. Yamanaka, *J. Alloys Compd.* **400**, 270 (2005).
176. Y. Miseki, H. Kato, A. Kudo, *Energy Environ. Sci.* **1062**, 306 (2009).
177. H. G. Kim, D. W. Hwang, J. Kim, Y. G. Kim, J. S. Lee, *Chem. Commun.* **12**, 1077 (1999).
178. T. Takata, Y. Furumi, K. Shinohara, A. Tanaka, M. Hara, J. N. Kondo, K. Domen, *Chem. Mater.* **9**, 1063 (1997).
179. Y. Huang, J. Wu, Y. Wei, S. Hao, M. Huang, J. Lin, *Scripta Mater.* **57**, 437 (2007).
180. T. Takata, A. Tanaka, M. Hara, J. N. Kondo, K. Domen, *Stud. Surf. Sci. Catal.* **130**, 1943 (2000).
181. Y. W. Tai, J. S. Chen, C. C. Yang, B. Z. Wan, *Catal. Today* **97**, 95 (2004).

182. T. Takata, K. Shinohara, A. Tanaka, M. Hara, J. N. Kondo, K. J. Domen, *Photochem. Photobiol. A: Chem.* **106**, 45 (1997).
183. Y. Yang, Q. Y. Chen, Z. L. Y, J. Li, *Appl. Surf. Sci.* **255**, 8419 (2009).
184. Y. Huang, J. Wu, Y. Wei, J. Lin, M. J. Huang, *Alloys Compd.* **456**, 364 (2008).
185. H. Takahashi, M. Kakihan, Y. Yamashita, K. Yoshida, S. Ikeda, M. Hara, K. J. Domen, *Alloys Compd.* **285**, 77 (1999).
186. V. R. Reddy, D. W. Hwang, J. S. Lee, *Catal. Lett.* **90**, 39 (2003).
187. T. Sekine, J. Yoshimura, A. Tanaka, K. Domen, K. Maruya, T. Onishi, *Bull. Chem. Soc. Jpn.* **63**, 2107 (1990).
188. A. Kudo, S. Hijii, *Chem. Lett.* **28**, 1103 (1999).
189. C. H. He, O. B. Yang, *Ind. Eng. Chem. Res.* **42**, 419 (2003).
190. K. Sayama and H. Arakawa, *J. Phys. Chem.* **97**, 531 (1993).
191. K. Sayarna, H. Arakawa, *J. Photochem. Photobiol. A: Chem.* **94**, 67 (1996).
192. V. R. Reddy, D. W. Hwang, J. S. Lee, *Korean J. Chem. Eng.* **20**, 1026 (2003).
193. J. J. Zou, C. J. Liu, Y. P. Zhang, *Langmuir* **22**, 2334 (2006).
194. S. H. Liu, H. P. Wang, *Int. J. Hydrogen Energy* **27**, 859 (2002).
195. Y. Yuan, X. Zhang, L. Liu, X. Jiang, J. Lv, Z. Li, Z. Zou, *Int. J. Hydrogen Energy* **33**, 5941 (2008).
196. Y. P. Yuan, Z. Y. Zhao, J. Zheng, M. Yang, L. G. Qiu, Z. S. Li, Z. G. J. Zou, *Mater. Chem.* **20**, 6772 (2010).
197. M. Uno, A. Kosuga, M. Okui, K. Horisaka, H. Muta, K. Kurosaki, S. J. Yamanaka, *Alloys Compd.* **420**, 291 (2006).
198. M. P. Kapoor, S. Inagaki, H. Yoshida, *J. Phys. Chem.* B **109**, 9231 (2005).
199. H. Byrd, A. Clearfield, D. Poojary, K. P. Reis, M. E. Thompson, *Chem. Mater.* **8**, 2239 (1996).
200. X. Chen, T. Yu, X. Fan, H. Zhang, Z. Li, J. Ye, Z. Zou, *Appl. Surf. Sci.* **253**, 8500 (2007).
201. H. Y. Lin, H. C. Huang, W. L. Wang, *Micropor. Mesopor. Mater.* **115**, 568 (2008).
202. K. Domen, A. Kudo, M. Shibata, A. Tanaka, K. Maruya, T. Onishi, *J. Chem. Soc. Chem. Commun.* **23**, 1706 (1986).
203. K. Domen, A. Kudo, A. Shinozaki, A. Tanaka, K. Maruya, T. Onishi, *J. Chem. Soc. Chem. Commun.* **4**, 356 (1986).
204. A. Kudo, A. Tanaka, K. Domen, K. Maruya, K. Aika, T. Onishi, *J. Catal.* **111**, 67 (1988).
205. S. Ikeda, A. Tanaka, K. Shinohara, M. Hara, J. N. Kondo, K. Maruya, K. Domen, *Micropor. Mater.* **9**, 253 (1997).

206. A. Kudo, K. Sayama, A. Tanaka, K. Asakura, K. Domen, K. Maruya, T. Onishi, *J. Catal.* **120**, 337 (1989).
207. K. Domen, A. Kudo, A. Tanaka, T. Onishi, *Catal. Today* **8**, 77 (1990).
208. K. Sayama, A. Tanaka, K. Domen, K. Maruya, T. Onishi, *Catal. Lett.* **4**, 217 (1990).
209. K. Sayama, A. Tanaka, K. Domen, K. Maruya, T. Onishi, *J. Phys. Chem.* **95**, 1345 (1991).
210. K. Sayama, K. Yase, H. Arakawa, K. Asakura, A. Tanaka, K. Domen, T. Onishi, *J. Photochem. Photobiol.* A: Chem. **114**, 125 (1998).
211. K. H. Chung, D. C. Park, *J. Mol. Catal. A: Chem.* **129**, 53 (1998).
212. K. Sayama, A. Tanaka, K. Domen, K. Maruya, T. Onishi, *J. Catal.* **124**, 541 (1990).
213. G. Li, T. Kako, D. Wang, Z. Zou, J. Ye, *J. Phys. Chem. Solids* **69**, 2487 (2008).
214. Q. P. Ding, Y. P. Yuan, X. Xiong, R. P. Li, H. B. Huang, Z. S. Li, T. Yu, Z. G. Zou, S. G. Yang, *J. Phys. Chem.* C **112**, 18846 (2008).
215. B. Zielinska, E. Borowiak-Palen, R. J. Kalenzuk, *J. Phys. Chem. Solids* **69**, 236 (2008).
216. Y. Miseki, H. Kato, A. Kudo, *Chem. Lett.* **34**, 54 (2005).
217. S. Ikeda, T. Itani, K. Nango, M. Matsumura, *Catal. Lett.* **98**, 229 (2004).
218. D. Chen, Ye, *J. Chem. Mater.* **21**, 2327 (2009).
219. A. Kudo, H. Kato and S. Nakagawa, *J. Phys. Chem.* B **104**, 571 (2000).
220. J. Yin, Z. Zou, J. Ye, *J. Phys. Chem.* B **108**, 8888 (2004).
221. J. Yin, Z. Zou, J. Ye, *J. Phys. Chem.* B **108**, 12790 (2004).
222. K. Domen, J. Yoshimura, T. Sekine, A. Tanaka, T. Onishi, *Catal. Lett.* **4**, 339 (1990).
223. Y. Li, J. Wu, Y. Huang, M. Huang, J. Lin, *Int. J. Hydrogen Energy* **34**, 7927 (2009).
224. Y. Huang, Y. Xie, L. Fan, Y. Li, Y. Wei, J. Lin, J. Wu, *Int. J. Hydrogen Energy* **33**, 6432 (2008).
225. Y. Wei, J. Li, Y. Huang, M. Huang, J. Lin, Wu, *J. Sol. Energy Mater. Sol. Cells* **93**, 1176 (2009).
226. Y. Huang, J. Li, Y. Wei, Y. Li, J. Lin, J. Wu, *J. Hazard. Mater.* **166**, 103 (2009).
227. Y. Huang, Y. Wei, L. Fan, M. Huang, J. Lin, J. Wu, *Int. J. Hydrogen Energy* **34**, 5318 (2009).
228. Y. Ebina, T. Sasaki, M. Harada, Watanabe, Mamoru. *Chem. Mater.* **14**, 4390 (2002).

229. Y. Ebina, N. Sakai, T. Sasaki, *J. Phys. Chem.* B **109**, 17212 (2005).
230. O. C. Compton, E. C. Carroll, J. Y. Kim, D. S. Larsen, F. E. Osterloh, *J. Phys. Chem. C* **111**, 14589 (2007).
231. Y. Ebina, A. Tanaka, J. N. Kondo, K. Domen, *Chem. Mater.* **8**, 2534 (1996).
232. K. Domen, Y. Ebina, T. Sekine, A. Tanaka, J. Kondo, C. Hirose, *Catal. Today* **16**, 479 (1993).
233. Y. Li, J. Wu, Y. Huang, M. Huang, *J. Alloys Compd.* **453**, 437 (2008).
234. R. Abe, M. Higashi, Z. Zou, K. Sayama, Y. Abe, H. Arakawa, *J. Phys. Chem. B* **108**, 811 (2004).
235. D. Li, J. Zheng, Z. Li, X. Fan, L. Liu, Z. Zou, *Int. J. Photoenergy*, 21860 (2007)
236. A. Kudo, S. Nakagawa, H. Kato, *Chem. Lett.* **28**, 1197 (1999).
237. Y. Li, G. Chen, H. Zhang, Z. Lv, *Int. J. Hydrogen Energy* **35**, 2652 (2010).
238. J. Luan, S. Zheng, X. Hao, G. Luan, X. Wu, Z. J. Zou, *Braz. Chem. Soc.* **17**, 1368 (2006).
239. Z. Zou, J. Ye, H. Arakawa, *Chem. Phys. Lett.* **333**, 57 (2001).
240. Y. Li, G. Chen, H. Zhang, Z. Li, *Mater. Res. Bull.* **44**, 741 (2009).
241. Z. Zou, J. Ye, H. Arakawa, *Top. Catal.* **22**, 107 (2003).
242. Z. Zou, J. Ye, K. Sayama, H. Arakawa, *Chem. Phys. Lett.* **343**, 303 (2001).
243. K. Sayama, H. Arakawa, K. Domen, *Catal. Today* **28**, 175 (1996).
244. H. Kato, A. Kudo, *Chem. Phys. Lett.* **295**, 487 (1998).
245. Y. Takahara, J. N. Kondo, T. Takata, D. Lu, K. Domen, *Chem. Mater.* **13**, 1194 (2001).
246. M. Stodolny, M. Laniecki, *Catal. Today* **142**, 314 (2009).
247. J. N. Kondo, M. Uchida, K. Nakajima, D. Lu, M. Hara, K. Domen, *Chem. Mater.* **16**, 4304 (2004).
248. Y. Takahara, J. N. Kondo, D. Lu, K. Domen, *Solid State Ionics* **151**, 305 (2002).
249. H. Kato, A. Kudo, *J. Phys. Chem.* B **105**, 4285 (2001).
250. H. Kato, A. Kudo, *Catal. Lett.* **58**, 153 (1999).
251. H. Kato, A. Kudo, *Catal. Today* **78**, 561 (2003).
252. H. Kato, A. Kudo, *Chem. Lett.* **28**, 1207 (1999).
253. Y. Lee, T. Watanabe, T. Takata, M. Hara, M. Yoshimura, K. Domen, *Bull. Chem. Soc. Jpn.* **80**, 423 (2007).
254. J. W. Liu, G. Chen, Z. H. Li, Z. G. Zhang, *Int. J. Hydrogen Energy* **32**, 2269 (2007).
255. C. C. Hu, H. Teng, Appl. Cata. A: *Gen.* **331**, 44 (2007).
256. C. Mitsui, H. Nishiguchi, K. Fukamachi, T. Ishihara, Y. Takita, *Chem. Lett.* **28**, 1327 (1999).

257. T. Ishihara, H. Nishiguchi, K. Fukamachi, Y. Takita, *J. Phys. Chem. B* **103**, 1 (1999).
258. A. Kudo and H. Kato, *Chem. Phys. Lett.* **331**, 373 (2000).
259. A. Iwase, H. Kato, H. Okutomi, A. Kudo, *Chem. Lett.* **33**, 1260 (2004).
260. H. Kato, K. Asakura and A. Kudo, *J. Am. Chem. Soc.* **125**, 3082 (2003).
261. A. Yamakata, T. Ishibashi, H. Kato, A. Kudo, H. Onishi, *J. Phys. Chem. B* **107**, 14383 (2003).
262. M. Maruyama, A. Iwase, H. Kato, A. Kudo, H. Onishi, *J. Phys. Chem. C* **113**, 13918 (2009).
263. S. Ikeda, M. Fubuki, Y. K. Takahara, M. Matsumura, Appl. Catal. A: *Gen.* **300**, 186 (2006).
264. T. Ishihara, N. S. Baik, N. Ono, H. Nishiguchi, Y. Takita, *J. Photochem. Photobiol. A: Chem.* **167**, 149 (2004).
265. K. Yoshioka, V. Petrykin, M. Kakihana, H. Kato, A. Kudo, *J. Catal.* **232**, 102 (2005).
266. M. Yoshino, M. Kakihana, W. S. Cho, H. Kato and A. Kudo, *Chem. Mater.* **14**, 3369 (2002).
267. H. Kato, A. Kudo, *J. Photochem. Photobiol. A: Chem.* **145**, 129 (2001).
268. H. Otsuka, K. Kim, A. Kouzu, I. Takimoto, H. Fujimori, Y. Sakata, H. Imamura, T. Matsumoto, K. Toda, *Chem. Lett.* **34**, 822 (2005).
269. B. Xu, W. F. Zhang, X. Y. Liu, J. H. Ye, W. H. Zhang, L. Shi, X. G. Wan, J. Yin, Z. G. Liu, *Phys. Rev. B* **76**, 125109 (2007).
270. K. Shimizu, Y. Tsuji, T. Hatamachi, K. Toda, T. Kodama, M. Sato, Y. Kitayama, Phys. *Chem. Chem. Phys.* **6**, 1064 (2004).
271. K. Shimizu, S. Itoh, T. Hatamachi, T. Kodama, M. Sato, K. Toda, *Chem. Mater.* **17**, 5161 (2005).
272. W. Yao, Ye, *J. Chem. Phys. Lett.* **435**, 96 (2007).
273. T. Mitsuyama, A. Tsutsumi, T. Hata, K. Ikeue, M. Machida, *Bull. Chem. Soc. Jpn.* **81**, 401 (2008).
274. Y. Li, G. Chen, C. Zhou, Z. Li, *Catal. Lett.* **123**, 80 (2008).
275. M. Machida, K. Miyazaki, S. Matsushima, M. Arai, *J. Mater. Chem.* **13**, 1433 (2003).
276. M. Machida, J. Yabunaka, T. Kijima, *Chem. Commun.* **19**, 1939 (1999).
277. M. Machida, J. Yabunaka, T. Kijima, *Chem. Mater.* **12**, 812 (2000).
278. M. Machida, J. Yabunaka, T. Kijima, S. Matsushima, M. Arai, *Int. J. Inorg. Mater.* **3**, 545 (2001).

279. A. Kudo, H. Okutomi, H. Kato, *Chem. Lett.* **29**, 1212 (2000).
280. M. Machida, S. Murakami, T. Kijima, S. Matsushima, M. Arai, *J. Phys. Chem. B* **105**, 3289 (2001).
281. A. Kudo and H. Kato, *Chem. Lett.* **26**, 867 (1997).
282. T. Kurihara, H. Okutomi, Y. Miseki, H. Kato, A. Kudo, *Chem. Lett.* **35**, 274 (2006).
283. J. F. Luan, X. P. Hao, S. R. Zheng, G. Y. Luan, X. S. Wu, *J. Mater. Sci.* **41**, 8001 (2006).
284. Y. Li, G. Chen, H. Zhang, Z. Li, *J. Phys. Chem. Solids* **70**, 536 (2009).
285. Y. Li, G. Chen, H. Zhang, Z. Li, J. Sun, *J. Solid State Chem.* **181**, 2653 (2008).
286. H. Kato, H. Kobayashi, A. Kudo, *J. Phys. Chem.* B **106**, 12441 (2002).
287. Y. Hosogi, K. Tanabe, H. Kato, H. Kobayashi, A. Kudo, *Chem. Lett.* **33**, 28 (2004).
288. H. Kadowaki, N. Saito, H. Nishiyama, H. Kobayashi, Y. Shimodaira, Y. Inoue, *J. Phys. Chem. C* **111**, 439 (2007).
289. N. Saito, H. Kadowaki, H. Kobayashi, K. Ikarashi, H. Nishiyama, Y. Inoue, *Chem. Lett.* **33**, 1452 (2004).
290. A. Kudo, A. Steinberg, A. J. Bard, A. Campion, M. A. Fox, T. E. Mallouk, S. E. Webber, J. M. White, *Catal. Lett.* **5**, 61 (1990).
291. H. Kato, N. Matsudo, A. Kudo, *Chem. Lett.* **33**, 1216 (2004).
292. T. V. Nguyen, K. J. Kim, O. B. Yang, *J. Photochem. Photobiol. A: Chem.* **173**, 56 (2005).
293. Y. Wang, Z. Zhang, Y. Zhu, Z. Li, R. Vajtai, L. Ci, P. M. Ajayan, *ACS Nano* **2**, 1492 (2008).
294. J. Luan, H. Cai, S. Zheng, X. Hao, G. Luan, X. Wu, Z. Zou, *Mater. Chem. Phys.* **104**, 119 (2007).
295. J. Luan, Z. Zheng, H. Cai, X. Wu, G. Luan, Z. Zou, *Mater. Res. Bull.* **43**, 3332 (2008).
296. T. Yanagida, Y. Sakata, H. Imamura, *Chem. Lett.* **33**, 726 (2004).
297. Y. Sakata, Y. Matsuda, T. Yanagida, K. Hirata, H. Imamura, K. Teramura, *Catal. Lett.* **125**, 22 (2008).
298. D. Wang, Z. Zou, Ye, *J. Chem. Phys. Lett.* **384**, 139 (2004).
299. A. Kudo, I. Mikami, *J. Chem. Soc. Faraday Trans.* **94**, 2929 (1998).
300. N. Arai, N. Saito, H. Nishiyama, Y. Shimodaira, H. Kobayashi, Y. Inoue, K. Sato, *J. Phys. Chem. C* **112**, 5000 (2008).
301. J. Sato, H. Kobayashi, S. Saito, H. Nishiyama, Y. Inoue, *J. Photochem. Photobiol. A: Chem.* **158**, 139 (2003).
302. J. Sato, S. Saito, H. Nishiyama, Y. Inoue, *J. Phys. Chem. B* **107**, 7965 (2003).

303. J. Sato, H. Kobayashi, Y. Inoue, *J. Phys. Chem. B* **107**, 7970 (2003).
304. J. Sato, N. Saito, H. Nishiyama, Y. Inoue, *J. Phys. Chem. B* **105**, 6061 (2001).
305. J. Sato, N. Saito, H. Nishiyama, Y. Inoue, *Chem. Lett.* **30**, 868 (2001).
306. J. Sato, S. Saito, H. Nishiyama, Y. Inoue, *J. Photochem. Photobiol. A: Chem.* **148**, 85 (2002).
307. K. Ikarashi, J. Sato, H. Kobayashi, N. Saito, H. Nishiyama, Y. Inoue, *J. Phys. Chem. B* **106**, 9048 (2002).
308. J. Sato, H. Kobayashi, K. Ikarashi, N. Saito, H. Nishiyama, Y. Inoue, *J. Phys. Chem. B* **108**, 4369 (2004).
309. H. Kadowaki, J. Sato, H. Kobayashi, N. Saito, H. Nishiyama, Y. Simodaira, Y. Inoue, *J. Phys. Chem. B* **109**, 22995 (2005).
310. W. F. Zhang, J. W. Tang, J. H. Ye, *Chem. Phys. Lett.* **418**, 174 (2006).
311. D. Chen, J. Ye, *Chem. Mater.* **19**, 4585 (2007).
312. G. R. Bamwenda, T. Uesigi, Y. Abe, K. Sayama, H. Arakawa, *Appl. Catal. A: Gen.* **205**, 117 (2001).
313. H. Kadowaki, N. Saito, H. Nishiyama, Y. Inoue, *Chem. Lett.* **36**, 440 (2007).
314. J. K. Reddy, G. Suresh, C. H. Hymavathi, V. D. Kumari, M. Subrahmanyam, *Catal. Today* **141**, 89 (2009).
315. Y. Yuan, J. Zheng, X. Zhang, Z. Li, T. Yu, J. Ye, Z. Zou, *Solid State Ionics* **178**, 1711 (2008).
316. A. F. Wells, "Structure Inorganic Chemistry", 4th ed., Clarendon Press, London, 1975.
317. "Transition Metal Oxides: Surface Chemistry and Catalysis", edited by H. H. Kung, Elsevier Science Publishers B. V., Amsterdam, The Netherlands (1989).
318. G. Pfaff, P. Reynders, *Chem. Rev.* **99**, 1963 (1999).
319. A. Salvador, M. C. Pascual-Marti, J. R. Adell, A. Requeni, J. G. March, *J. Pharm. Biomed. Anal.* **22**, 301 (2000).
320. R. Zallen, M. P. Moret, *Solid State Commun.* **137**, 154 (2006).
321. J. H. Braun, A. Baidins, R. E. Marganski, *Prog. Org. Coat.* **20**, 105 (1992).
322. S. A. Yuan, W. H. Chen, S. S. Hu, *Mater. Sci. Eng. C* **25**, 479 (2005).
323. D. A. Tryk, A. Fujishima, K. Honda, *Electrochim. Acta* **45**, 2363 (2000).
324. M. Grätzel, Photoelectrochemical cells. *Nature* **414**, 338–344 (2001).

325. A. Hagfeldt, M. Grätzel, *Chem. Rev.* **95**, 49 (1995).
326. A. L. Linsebigler, G. Lu, J. T. Jr. Yates, *Chem. Rev.* **95**, 735 (1995).
327. A. Millis, S. Le Hunte, *J. Photochem. Photobiol. A* **108**, 1 (1997).
328. Xiaobo Chen and Samuel S. Mao, *Chem. Rev.* **107**, 2891–2959 (2007).
329. B. O. Regan and M. Grätzel, *Nature* **353**, 737–740 (1991).
330. M. Lazzeri, A. Vittadini, and A. Selloni, *Phys. Rev. B* **63**, 155409 (2001).
331. X. Gong and A. Selloni, *J. Phys. Chem.* B **109**, 19560–19562 (2005).
332. G. S. Herman, M. R. Sievers, and Y. Gao, *Phys. Rev. Lett.* **84**, 3354–3357 (2000).
333. A. Vittadini, A. Selloni, F. P. Rotzinger, and M. Grätzel, *Phys. Rev. Lett.* 8**1**, 2954–2957 (1998).
334. A. Vittadini, M. Casarin, and A. Selloni, *Theor. Chem. Acc.* **117**, 663–671 (2007).
335. M. Lazzeri and A. Selloni, *Phys. Rev. Lett.* **87**, 266105 (2001).
336. R. L. Penn and J. F. Banfield, *Geochim. Cosmochim. Acta* **63**, 15491557 (1999).
337. A. Zaban, S. T. Aruna, S. Tirosh, B. A. Gregg, and Y. Mastai, *J. Phys. Chem. B* **104**, 4130–4133 (2000).
338. Y. W. Jun, et al. *J. Am. Chem. Soc.* **125**, 15981–15985 (2003).
339. H. Zhang and J. F. Banfield, *J. Mater. Chem.* **8**, 2073 (1998).
340. H. Zhang and J. F. Banfield, *J. Phys. Chem. B* **104**, 3481 (2000).
341. Y. Hwu, Y. D. Yao, N. F. Cheng, C. Y. Tung, and H. M. Lin, *Nanostruct. Mater.* **9**, 355 (1997).
342. A. A. Gribb and J. F. Banfield, *Am. Mineral.* **82**, 717 (1997).
343. X. Ye, J. Sha, Z. Jiao, and L. Zhang, *Nanostruct. Mater.* **8**, 919 (1998).
344. H. Kominami, M. Kohno, Y. Kera, *J. Mater. Chem.* **10**, 1151 (2000).
345. M. R. Ranade, A. Navrotsky, H. Z. Zhang, J. F. Banfield, S. H. Elder, A. Zaban, P. H. Borse, S. K. Kulkarni, G. S. Doran, H. J. Whitfield, *Proc. Natl. Acad. Sci.* **99**, 6476 (2002).
346. W. Li, C. Ni, H. Lin, C. P. Huang, S. I. Shah, *J. Appl. Phys.* **96**, 6663 (2004).
347. R. W. G. Wyckoff, 1963 *Crystal Structures* 2nd edn, vol 1 and 2 (New York: Interscience).
348. R. Asahi, Y. Taga, W. Mannstadt, A. Freeman, *J. Phys. ReV. B* **61**, 7459 (2000).
349. P. I. Sorantin, K. Schwarz, *Inorg. Chem.* **31**, 567 (1992).

350. Z. Y. Wu, G. Ouvrared, P. Gressier, C. R. Natoli, *Phys. ReV. B* **55**, 10382 (1997).
351. R. Brydson, B. G. Williams, W. Engel, H. Sauer, E. Zeitler, J. M. Thomas, *Solid State Commun.* **64**, 609 (1987).
352. R. Brydson, H. Sauer, W. Engel, J. M. Thomas, E. Zeitler, N. Kosugi, H. J. Kuroda, *Phys.: Condens. Matter* **1**, 797 (1989).
353. H. C. Choi, H. J. Ahn, Y. M. Jung, M. K. Lee, H. J. Shin, S. B. Kim, Y. E. Sung, *Appl. Spectrosc.* **58**, 598 (2004).
354. L. D. Finkelstein, E. Z. Kurmaev, M. A. Korotin, A. Moewes, B. Schneider, S. M. Butorin, J.-H. Guo, J. Nordgren, D. Hartmann, M. Neumann, D. L. Ederer, *Phys. ReV. B* **60**, 2212 (1999).
355. V. Luca, S. Djajanti, R. F. Howe, *J. Phys. Chem. B* **102**, 10650 (1998).
356. R. Sanjines, H. Tang, H. Berger, F. Gozzo, G. Margaritondo, F. Levy, *J. Appl. Phys.* **75**, 2945 (1994).
357. R. Zimmermann, P. Steiner, R. Claessen, F. Reinert, S. Hufner, *J. Electron. Spectrosc. Relat. Phenom.* **96**, 179 (1998).
358. Jinghua Guo, Per-Anders Glans, Yi-Sheng Liu, and Chinglin Chang, Electronic Structure Study of Nanostructured Transition Metal Oxides Using Soft X-Ray Spectroscopy, Chapter 5 In *On Solar Hydrogen and Nanotechnology*, edited by Lionel Vayssieres, John Wiley and Sons (Asia) Pte Ltd. (2009).
359. P. Blaha, K. Schwarz, G. K. H. Madsen, D. Kvasnicka, and J. Luitz, *WIEN2k, An Augmented Plane Wave+Local Orbitals Program for Calculating Crystal Properties* (Karlheinz Schwarz, Techn. Univ. Wien, Austria, 2001).
360. J. P. Perdew and Y. Wang, *Phys. Rev. B* **45**, 13 244 (1992).
361. V. I. Anisimov, I. V. Solovyev, M. A. Korotin, M. T. Czyzyk, and G. A. Sawatzky, *Phys. Rev. B* **48**, 16 929 (1993).
362. A. I. Liechtenstein, V. I. Anisimov, and J. Zaanen, *Phys. Rev. B* **52**, R5467 (1995).
363. P. Novák, F. Boucher, P. Gressier, P. Blaha, and K. Schwarz, *Phys. Rev. B* **63**, 235114 (2001).
364. C. Persson and A. Zunger, *Phys. Rev. B* **68**, 073205 (2003).
365. S. Massidda, R. Resta, M. Posternak, and A. Baldereschi, *Phys. Rev. B* **52**, 16 977 (1995).
366. N. Sakai, Y. Ebina, K. Takada, T. Sasaki, *J. Am. Chem. Soc.* **126**, 5851 (2004).
367. T. Sasaki, M. Watanabe, *J. Phys. Chem. B* **101**, 10159 (1997).
368. L. Braginsky, V. Shklover, *Eur. Phys. J. D* **9**, 627 (1999).

369. H. Sato, K. Ono, T. Sasaki, A. Yamagishi, *J. Phys. Chem. B* **107**, 9824 (2003).
370. T. Sasaki, T. Supramol. *Sci.* **5**, 367 (1998).
371. D. V. Bavykin, S. N. Gordeev, A. V. Moskalenko, A. A. Lapkin, F. C. Walsh, *J. Phys. Chem. B* **109**, 8565 (2005).
372. C. Kormann, D. W. Bahnemann, and M. R. Hoffmann, J. *Phys. Chem.* **92**, 5196 (1988).
373. J. B. Goodenough and A. Hamnett, in *Landolt-Bernstein*, New Series, Group III, Vol. 17g, ed. O. Madelung, Springer Verlag, Berlin, Heidelberg, New York, Tokyo, 1984, p. 147.
374. L. E. Brus, *J. Phys. Chem.* **90**, 2555 (1986).
375. X. Chen, Y. Lou, A. C. S. Samia, C. Burda, *Nano Lett.* **3**, 799 (2003).
376. C. Burda, Y. Lou, X. Chen, A. C. S. Samia, J. Stout, J. L. Gole, *Nano Lett.* **3**, 1049 (2003).
377. M. Matsumura, Y. Saho, H. Tsubomura, *J. Phys. Chem.* **87**, 3807 (1983).
378. J. F. Reber, K. Meier, *J. Phys. Chem.* **90**, 824 (1986).
379. J. R. Darwent, A. Mills, *J. Chem. Soc. Faraday Trans. 2: Mol. Chem. Phys.* **78**, 359 (1982).
380. W. Erbs, J. Desilvestro, E. Borgarello, M. Grätzel, *J. Phys. Chem.* **88**, 4001 (1984).
381. Y. Miseki, H. Kusama, H. Sugihara, K. Sayama, *J. Phys. Chem. Lett.* **1**, 1196 (2010).
382. D. Meissner, R. Memming, B. Kastening, *J. Phys. Chem.* **92**, 3476 (1988).
383. D. Meissner, R. Memming, B. Kastening, D. Bahnemann, *Chem. Phys. Lett.* **127**, 419–423 (1986).
384. A. J. Frank, K. Honda, *J. Phys. Chem.* **86**, 1933 (1982).
385. H. Zhang, Y. F. Zhu, *J. Phys. Chem. C* **114**, 5822 (2010).
386. X. X. Yan, G. Liu, L. Z. Wang, Y. Wang, X. F. Zhu, J. Zou, G. Q. Lu, *J. Mater. Res.* **25**, 182 (2010).
387. T. Torimoto, M. Hashitani, T. Konishi, K. I. Okazaki, T. Shibayama, B. Ohtani, *J. Nanosci. Nanotechnol.* **9**, 506 (2009).
388. R. Solarska, B. D. Alexander, A. Braun, R. Jurczakowski, G. Fortunato, M. Stiefel, T. Graule, Augustynski, *J. Electrochim. Acta* **55**, 7780–7787 (2010).
389. N. Gaillard, B. Cole, J. Kaneshiro, E. L. Miller, B. Marsen, L. Weinhardt, M. Br, C. Heske, K. S. Ahn, Y. Yan, M. M. Al-Jassim, *J. Mater. Res.* **25**, 45 (2010).
390. M. Bär, L. Weinhardt, B. Marsen, B. Cole, N. Gaillard, E. Miller, C. Heske, *Appl. Phys. Lett.* **96**, 032107 (2010).

391. J. Augustynski, R. Solarska, H. Hagemann, C. Santato, *Proc. SPIE* **6340**, U140 (2006).
392. R. Solarska, B. D. Alexander, J. Augustynski, *CR Chim.* **9**, 301 (2006).
393. R. Solarska, C. Santato, C. Jorand-Sartoretti, M. Ulmann, J. Augustynski, *J. Appl. Electrochem.* **7**, 715 (2005).
394. R. Solarska, B. D. Alexander, J. Augustynski, *J. Solid State Electrochem.* **8**, 748 (2004).
395. X. Chen, S. Mao, *S. Chem. Rev.* **107**, 2891 (2007).
396. P. F. Ji, M. Takeuchi, T. M. Cuong, J. L. Zhang, M. Matsuoka, M. Anpo, *Res. Chem. Intermediat.* **36**, 327 (2010).
397. D. Y. C. Leung, X. L. Fu, C. F. Wang, M. Ni, M. K. H. Leung, X. X. Wang, X. Z. Fu, *ChemSusChem* **3**, 681 (2010).
398. H. Kato, A. Kudo, *J. Phys. Chem.* B **106**, 5029 (2002).
399. J. Liu, G. Chen, Z. Lia, Z. Zhang, *J. Solid State Chem.* **179**, 3704 (2006).
400. D. W. Hwang, H. G. Kim, J. S. Lee, W. Li, S. H. Oh, *J. Phys. Chem. B* **109**, 2093 (2005).
401. A. Kudo, M. Sekizawa, *Catal. Lett.* **58**, 241 (1999).
402. A. Kudo, M. Sekizawa, *Chem. Commun.* **15**, 1371 (2000).
403. E. Borgarello, J. Kiwi, M. Grätzel, E. Pelizzetti, M. Visca, *J. Am. Chem. Soc.* **104**, 2996 (1982).
404. Z. Luo, Q. Gao, *J. Photochem. Photobiol. A: Chem.* **63**, 367 (1992).
405. B. Tian, C. Li, F. Gu, H. Jiang, Y. Hu, Zhang, *J. Chem. Eng. J.* **151**, 220 (2009).
406. L. G. Devi, S. G. Kumar, B. N. Murthy, N. Kottam, *Catal. Commun.* **10**, 794 (2009).
407. G. H. Takaoka, T. Nose, M. Kawashita, *Vacuum* **83**, 679 (2008).
408. X. Fan, X. Chen, S. Zhu, Z. Li, T. Yu, J. Ye, Z. Zou, *J. Mol. Catal. A: Chem.* **284**, 155 (2008).
409. D. H. Kim, D. K. Choi, S. J. Kim, K. S. Lee, *Catal. Commun.* **9**, 654 (2008).
410. T. Umebayashi, T. Yamaki, H. Itoh, K. Asai, *J. Phys. Chem. Solids* **63**, 1909 (2002).
411. T. Nishikawa, Y. Shinohara, T. Nakajima, M. Fujita, S. Mishima, *Chem. Lett.* **28**, 1133 (1999).
412. Y. Cao, W. Yang, W. Zhang, G. Liu, P. Yue, *New J. Chem.* **28**, 218 (2004).
413. S. Klosek, D. Raftery, *J. Phys. Chem.* B **105**, 2815 (2001).
414. M. A. Khan, S. I. Woo, O. B. Yang, *Int. J. Hydrogen Energy* **33**, 5345 (2008).

415. R. Dholam, N. Patel, M. Adami, A. Miotello, *Int. J. Hydrogen Energy* **34**, 5337 (2009).
416. D. Eder, M. Motta, A. H. Windle, *Nanotech.* **20**, 055602 (2009).
417. M. Anpo, *Pure Appl. Chem.* **72**, 1787 (2000).
418. M. Anpo, *Pure Appl. Chem.* **72**, 1265 (2000).
419. M. Anpo, S. Kishiguchi, Y. Ichihashi, M. Takeuchi, H. Yamashita, K. Ikeue, B. Morin, A. Davidson, M. Che, *Res. Chem. Intermed.* **27**, 459 (2001).
420. M. Anpo, M. Takeuchi, *Int. J. Photoenergy* **3**, 89 (2001).
421. M. Anpo, M. Takeuchi, *J. Catal.* **216**, 505 (2003).
422. M. Takeuchi, H. Yamashita, M. Matsuoka, M. Anpo, T. Hirao, N. Itoh, N. Iwamoto, *Catal. Lett.* **67**, 135 (2000).
423. S. Kim, S. J. Hwang, W. Choi, *J. Phys. Chem. B* **109**, 24260 (2005).
424. S. Rengaraj, X. Z. Li, *J. Mol. Catal. A: Chem.* **243**, 60 (2006).
425. W. Y. Choi, A. Termin, M. R. J. Hoffmann, *Phys. Chem.* **84**, 13669 (1994).
426. M. Grätzel, R. F. Howe, *J. Phys. Chem.* **94**, 2566 (1990).
427. R. Niishiro, R. Konta, H. Kato, W. J. Chun, K. Asakura, A. Kudo, *J. Phys. Chem. C* **111**, 17420 (2007).
428. R. Niishiro, H. Kato, A. Kudo, *Phys. Chem. Chem. Phys.* **7**, 2241 (2005).
429. T. Ikeda, T. Nomoto, K. Eda, Y. Mizutani, H. Kato, A. Kudo, H. Onishi, *J. Phys. Chem. C* **112**, 1167 (2008).
430. Y. Matsumoto, U. Unal, N. Tanaka, A. Kudo, H. Kato, *J. Solid State Chem.* **177**, 4205 (2004).
431. T. Ohno, F. Tanigawa, K. Fujihara, S. Izumi, M. Matsumura, *J. Photochem. Photobiol. A: Chem.* **127**, 107 (1999).
432. M. A. Khan, O. B. Yang, *Catal. Today* **146**, 177 (2009).
433. M. A. Khan, M. S. Akhtar, S. I. Woo, O. B. Yang, *Catal. Commun.* **10**, 1 (2008).
434. R. Sasikala, V. Sudarsan, C. Sudakar, R. Naik, L. Panicker, S.R. Bharadwaj, *Int. J. Hydrogen Energy* **34**, 6105 (2009).
435. M. Kitano, M. Takeuchi, M. Matsuoka, J. M. Thomas, M. Anpo, *Catal. Today* **120**, 133 (2007).
436. R. Dholam, N. Patel, M. Adami, A. Miotello, *Int. J. Hydrogen Energy* **33**, 6896 (2008).
437. F. Zuo, L. Wang, T. Wu, Z. Zhang, D. Borchardt, P. Feng, *J. Am. Chem. Soc.* **132**, 1185611857 (2010).
438. D. Wang, J. Ye, T. Kako, T. Kimura, *J. Phys. Chem.* B **110**, 15824 (2006).
439. R. Konta, T. Ishii, H. Kato, A. Kudo, *J. Phys. Chem.* B **108**, 8992 (2004).

440. K. Sayama, K. Mukasa, R. Abe, Y. Abe, H. Arakawa, *Chem. Commun.* **23**, 2416 (2001).
441. S. Nishimoto, M. Matsuda, M. Miyake, *Chem. Lett.* **35**, 308 (2006).
442. H. Zhang, G. Chen, Y. Li, Y. Teng, *Int. J. Hydrogen Energy* **35**, 2713 (2010).
443. Y. Okazaki, T. Mishima, S. Nishimoto, M. Matsuda, M. Miyake, *Mater. Lett.* **62**, 3337 (2008).
444. D. W. Hwang, H. G. Kim, J. S. Jang, S. W. Bae, S. M. Ji, J. S. Lee, *Catal. Today* **93–95**, 845 (2004).
445. B. Wang, C. Li, D. Hirabayashi, K. Suzuki, *Int. J. Hydrogen Energy* **35**, 3306 (2010).
446. Y. Yang, Q. Chen, Z. Yin, J. Li, *J. Alloys Compd.* **225**, 8419 (2008).
447. P. Shah, D. S. Bhange, A. S. Deshpande, M. S. Kulkarni, N. M. Gupta, *Mater. Chem. Phys.* **117**, 399 (2009).
448. Z. Zou, J. Ye, K. Sayama, H. Arakawa, *Nature* **414**, 625 (2001).
449. Z. Zou, J. Ye, K. Sayama, H. Arakawa, *J. Photochem. Photobiol. A: Chem.* **148**, 65 (2002).
450. Z. Zou, J. Ye, H. Arakawa, *Catal. Lett.* **75**, 209 (2001).
451. Z. Zou, J. Ye, R. Abe, K. Sayama, H. Arakawa, *Stud. Surf. Sci. Catal.* **145**, 165 (2003).
452. Z. Zou, J. Ye, H. Arakawa, *J. Phys. Chem. B* **106**, 13098 (2002).
453. H. Y. Lin, T. H. Lee, C. Y. Sie, *Int. J. Hydrogen Energy* **33**, 4055 (2008).
454. H. Zhang, G. Chen, X. Li, Q. Wang, *Int. J. Hydrogen Energy* **34**, 3631 (2009).
455. H. Zhang, G. Chen, X. Li, *Solid State Ionics* **180**, 1599 (2009).
456. A. Iwase, K. Saito, A. Kudo, *Bull. Chem. Soc. Jpn.* **82**, 514 (2009).
457. M. Yang, X. Huang, S. Yan, Z. Li, T. Yu, Z. Zou, *Mater. Chem. Phys.* **121**, 506 (2010).
458. Y. Shimodaira, H. Kato, H. Kobayashi, A. Kudo, *Bull. Chem. Soc. Jpn.* **80**, 885 (2007).
459. N. Zeug, J. Bucheler, H. Kisch, *J. Am. Chem. Soc.* **107**, 1459 (1985).
460. J. H. Bang, R. J. Helmich, K. S. Suslick, *Adv. Mater.* **20**, 2599 (2008).
461. L. Ren, F. Yang, Y. R. Deng, N. N. Yan, S. Huang, D. Lei, Q. Sun, Y. Yu, *Int. J. Hydrogen Energy* **35**, 3297 (2010).
462. K. Ikeue, S. Shiiba, M. Machida, *Chem. Mater.* **22**, 743 (2010).
463. G. Liu, L. Zhao, L. Ma, L. Guo, *Catal. Commun.* **9**, 126 (2008).
464. W. Zhang, Z. Zhong, Y. Wang, R. Xu, *J. Phys. Chem. C* **112**, 17635 (2008).

465. Y. Wang, Y. Wang, R. Xu, *Int. J. Hydrogen Energy* **35**, 5245 (2010).
466. W. Zhang, R. Xu, *Int. J. Hydrogen Energy* **34**, 8495 (2009).
467. M. Subrahmanyam, V. T. Supriya, P. R. Reddy, *Int. J. Hydrogen Energy* **21**, 99 (1996).
468. A. M. Roy, G. C. De, *J. Photochem. Photobiol. A: Chem.* **157**, 87 (2003).
469. X. Zhang, D. Jing, M. Liu, L. Guo, *Catal. Commun.* **9**, 1720 (2008).
470. Z. Lei, W. You, M. Liu, G. Zhou, T. Takata, M. Hara, K. Domen; C. Li, *Chem. Commun.* **17**, 2142 (2003).
471. S. Shen, L. Zhao, L. Guo, *J. Phys. Chem. Solids* **69**, 2426 (2008).
472. S. Shen, L. Zhao, L. Guo, *Mater. Res. Bull.* **44**, 100 (2009).
473. S. Shen, L. Zhao, Z. Zhou, L. Guo, *J. Phys. Chem. C* **112**, 16148 (2008).
474. M. Ni, M. K. H. Leung, D. Y. C. Leung, K. Sumathy, *Renew. Sust. Energy Rev.* **11**, 401 (2007).
475. T. Lindgren, J. M. Mwabora, E. Avendano, J. Jonsson, A. Hoel, C. G. Granqvist, S. E. Lindquist, *J. Phys. Chem. B* **107**, 5709 (2003).
476. X. Chen, C. Burda, *J. Phys. Chem. B* **108**, 15446 (2004).
477. J. L. Gole, J. D. Stout, C. Burda, Y. Lou, X. Chen, *J. Phys. Chem. B* **108**, 1230 (2004).
478. A. Zaleska, E. Grabowska, J. W. Sobczak, M. Gazda, Hupka, *J. Appl. Catal. B: Environ.* **89**, 469 (2009).
479. J. Wang, S. Yin, M. Komatsu, T. Sato, *J. Eur. Ceram. Soc.* **25**, 3207 (2005).
480. T. Ohno, T. Tsubota, Y. Nakamura, K. Sayama, *App. Catal. A: Gen.* **288**, 74 (2005).
481. C. L. Paven-Thivet, A. Ishikawa, A. Ziani, L. L. Gendre, M. Yoshida, J. Kubota, F. Tessier, K. Domen, *J. Phys. Chem. C* **113**, 6156 (2009).
482. J. H. Yan, Y. R. Zhu, Y. G. Tang, S. Q. Zheng, *J Alloys Compd.* **472**, 429 (2009).
483. S. Ito, K. R. Thampi, P. Comte, P. Liska, M. Grätzel, *Chem. Commun.* **2**, 268 (2005).
484. D. Lu, G. Hitoki, E. Katou, J. N. Kondo, M. Hara, K. Domen, *Chem. Mater.* **16**, 1063 (2004).
485. M. Hara, G. Hitoki, T. Takata, J. N. Kondo, H. Kobayashi, K. Domen, *Catal. Today* **78**, 555 (2003).
486. M. Yashima, Y. Lee, K. Domen, *Chem. Mater.* **19**, 588 (2007).
487. G. Hitoki, T. Takata, J. N. Kondo, M. Hara, H. Kobayashi, K. Domen, *Electrochem.* **70**, 463 (2002).

488. M. Liu, W. You, Z. Lei, G. Zhou, J. Yang, G. Wu, G. Ma, G. Luan, T. Takata, M. Hara, K. Domen, C. Li, *Chem. Commun.* **19**, 2192 (2004).
489. M. Liu, W. You, Z. Lei, T. Takata, K. Domen, C. Li, J. Chin. *Catal.* **27**, 556 (2006).
490. M. Higashi, R. Abe, K. Teramura, T. Takata, B. Ohtani, K. Domen, *Chem. Phys. Lett.* **452**, 120 (2008).
491. K. G. Kanade, J. O. Baeg, B. B. Kale, S. M. Lee, S. J. Moon, K. Kong, *Int. J. Hydrogen* Energy **32**, 4678 (2007).
492. S. M. Ji, P. H. Borse, H. G. Kim, D. W. Hwang, J. S. Jang, S. W. Bae, J. S. Lee, *Phys. Chem. Chem. Phys.* 2005, 7, 1315.
493. H. Shi, X. Li, H. Iwai, Z. Zou, J. Ye, *J. Phys. Chem. Solids* **70**, 931 (2009).
494. Y. Matsumoto, M. Koinuma, Y. Iwanaga, T. Sato, S. Ida, *J. Am. Chem. Soc.* **131**, 6644 (2009).
495. Y. Sakatani, H. Ando, K. Okusako, H. Koike, J. Nunoshige, T. Takata, J. N. Kondo, M. Hara, K. Domen, *J. Mater. Res.* **19**, 2100 (2004).
496. Y. Choi, T. Umebayashi, M. Yoshikawa, *J. Mater. Sci.* **39**, 1837 (2004).
497. M. Shen, Z. Wu, H. Huang, Y. Du, Z. Zou, P. Yang, *Mater. Lett.* **60**, 693 (2006).
498. T. Ohno, T. Mitsui, M. Matsumura, *Chem. Lett.* **32**, 364 (2003).
499. T. Tesfamichael, G. Will, Bell, *J. Appl. Surf. Sci.* **245**, 172 (2005).
500. A. R. Gandhe, S. P. Naik, J. B. Fernandes, *Micropor. Mesopor. Mater.* **87**, 103 (2005).
501. Y. Zhao, X. Qiu, C. Burda, *Chem. Mater.* **20**, 2629 (2008).
502. G. Zhang, X. Ding, Y. Hu, B. Huang, X. Zhang, X. Qin, J. Zhou, J. Xie, *J. Phys. Chem. C* **112**, 17994 (2008).
503. J. Fang, F. Wang, K. Qian, H. Bao, Z. Jiang, W. Huang, *J. Phys. Chem. C* **112**, 18150 (2008).
504. D. Gu, Y. Lu, B. Yang, Y. Hu, *Chem. Commun.* **21**, 2453 (2008).
505. X. Chen, Y. Lou, A. C. S. Samia, C. Burda, J. L. Gole, *Adv. Funct. Mater.* **15**, 41 (2005).
506. Y. Liu, X. Chen, J. Li, C. Burda, *Chemosphere* **61**, 11 (2005).
507. G. Liu, C. Sun, X. Yan, L. Cheng, Z. Chen, X. Wang, L. Wang, S. C. Smith, G. Q. Lu, H. M. Cheng, *J. Mater. Chem.* **19**, 2822 (2009).
508. X. Chen, P. A. Glans, X. Qiu, S. Dayal, W. D. Jennings, K. E. Smith, C. Burda, J. Guo, *J. Electron. Spectrosc. Relat. Phenom.* **162**, 67 (2008).
509. X. Chen, C. Burda, *J. Am. Chem. Soc.* **130**, 5018 (2008).

510. R. Asahi, T. Morikawa, T. Ohwaki, K. Aoki, Y. Taga, *Science* **293**, 269 (2001).
511. A. Braun, K. K. Akurati, G. Fortunato, F. A. Reifler, A. Ritter, A. S. Harvey, A. Vital, T. Graule, *J. Phys. Chem. C* **114**, 516 (2010).
512. M. C. Yang, T. S. Yang, M. S. Wong, *Thin Solid Films* **469–470**, 1 (2004).
513. Y. Suda, H. I. Kawasak, T. I. Ueda, T. Ohshima, *Thin Solid Films* **453–454**, 162 (2004).
514. K. Kobayakawa, K. Murakami, Y. Sato, *J. Photochem. Photobiol. A: Chem.* **170**, 177 (2004).
515. S. Z. Chen, P. Y. Zhang, D. M. Zhuang, W. P. Zhu, *Catal. Commun.* **5**, 677 (2004).
516. G. R. Torres, T. Lindgren, J. Lu, C. G. Granqvist, S. E. Lindquist, *J. Phys. Chem. B* **108**, 5995 (2004).
517. Z. Jiang, F. Yang, N. Luo, B. T. T. Chu, D. Sun, H. Shi, T. Xiao, P. P. Edwards, *Chem. Commun.* **47**, 6372 (2008).
518. D. Huang, S. Liao, S. Quan, L. Liu, Z. He, J. Wan, W. Zhou, *J. Non-Cryst. Solids* **354**, 3965 (2008).
519. M. Mrowetz, W. Balcerski, A. J. Colussi, M. R. Hoffmann, *J. Phys. Chem.* B **108**, 17269 (2004).
520. L. Mi, P. Xu, P. N. Wang, *Appl. Surf. Sci.* **255**, 2574 (2008).
521. J. Yuan, M. Chen, J. Shi, W. Shangguan, *Int. J. Hydrogen Energy* **31**, 1326 (2006).
522. W. C. Lin, W. D. Yang, I. L. Huang, T. S. Wu, Z. Chung, *J. Energy Fuels* **23**, 2192 (2009).
523. S. C. Pillai, P. Periyat, R. George, D. E. McCormack, M. K. Seery, H. Hayden, J. Colreavy, D. Corr, S. J. Hinder, *J. Phys. Chem. C* **111**, 1605 (2007).
524. S. C. Padmanabhan, S. C. Pillai, J. Colreavy, S. Balakrishnan, D. E. McCormack, T. S. Perova, Y. Gunko, S. J. Hinder, J. M. Kelly, *Chem. Mater.* **19**, 4474 (2007).
525. P. Periyat, S. C. Pillai, D. E. McCormack, J. Colreavy, S. J. Hinder, *J. Phys. Chem. C* **112**, 7644 (2008).
526. T. Umebayashi, T. Yamaki, H. Itoh, K. Asai, *Appl. Phys. Lett.* **81**, 454 (2002).
527. T. Umebayashi, T. Yamaki, S. Tanala, K. Asai, *Chem. Lett.* **32**, 330 (2003).
528. J. C. Yu, W. K. Ho, J. G. Yu, H. Y. Yip, P. K. Wong, J. C. Zhao, *Environ. Sci. Technol.* **39**, 1175 (2005).
529. W. Ho, J. C. Yu, S. J. Lee, *Solid State Chem.* **179**, 1171 (2006).
530. K. Takeshita, A. Yamakata, T. A. Ishibashi, H. Onishi, K. Nishijima, T. Ohno, *J. Photochem. Photobiol. A: Chem.* **177**, 269 (2006).

531. K. Nishijima, T. Kamai, N. Murakami, T. Tsubota, T. Ohno, *Int. J. Photoenergy* 173943 (2008).
532. S. U. M. Khan, M. Al-Shahry, W. B. Ingler Jr., *Science* **297**, 2243 (2002).
533. C. Xu, Y. A. Shaban, W. B. Ingler Jr., S. U. M. Khan, *Sol. Energy Mater. Sol. Cells* **91**, 938 (2007).
534. Y. L. Su, X. W. Zhang, S. Han, X. Q. Chen, L. C. Lei, *Electrochem. Commun.* **9**, 2291 (2007).
535. E. A. Reyes-Garcia, Y. Sun, D. Raftery, *J. Phys. Chem. C* , **111**, 17146 (2007).
536. D. Li, H. Haneda, S. Hishita, N. Ohashi, *Chem. Mater.* **17**, 2588 (2005).
537. D. Li, H. Haneda, S. Hishita, N. Ohashi, *Chem. Mater.* **17**, 2596 (2005).
538. X. Chen, Y. Su, X. Zhang, L. Lei *Chin. Sci. Bull.* **53**, 1983 (2008).
539. M. Lim, Y. Zhou, B. Wood, Y. Guo, L. Wang, V. Rudolph, G. Lu, *J. Phys. Chem. C* **112**, 19655 (2008).
540. Y. Cong, F. Chen, J. L. Zhang, M. Anpo, *Chem. Lett.* **35**, 800 (2006).
541. J. Yang, H. Bai, Q. Jiang, J. Lian, *Thin Solid Films* **516**, 1736 (2008).
542. J. G. Yu, M. H. Zhou, B. Cheng, X. J. Zhao, *J. Mol. Catal. A: Chem.* **246**, 176 (2006).
543. P. Periyat, D. E. McCormack, S. J. Hinder, S. C. Pillai, *J. Phys. Chem. C* **113**, 3246 (2009).
544. Y. Sheng, Y. Xu, D. Jiang, L. Liang, D. Wu, Y. Sun, *Int. J. Photoenergy* 563949 (2007).
545. S. In, A. Orlov, R. Berg, F. Garca, S. Pedrosa-Jimenez, M. S. Tikhov, D. S. Wright, R. M. Lambert, *J. Am. Chem. Soc.* **129**, 13790 (2007).
546. N. Luo, Z. Jiang, H. Shi, F. Cao, T. Xiao, P. P. Edwards, *Int. J. Hydrogen Energy* **34**, 125 (2009).
547. H. Ozaki, S. Iwamoto, M. Inoue, *J. Mater. Sci.* **42**, 4009 (2007).
548. H. Sun, Y. Bai, Y. Cheng, W. Jin, N. Xu, *Ind. Eng. Chem. Res.* **45**, 4971 (2006).
549. X. Chen, X. Zhang, Y. Su, L. Lei, *Appl. Surf. Sci.* **254**, 6693 (2008).
550. H. Luo, T. Takata, Y. Lee, J. Zhao, K. Domen, Y. Yan, *Chem. Mater.* **16**, 846 (2004).
551. K. Nukumizu, J. Nunoshige, T. Takata, J. N. Kondo, M. Hara, H. Kobayashi, K. Domen, *Chem. Lett.* **32**, 196 (2003).
552. K. Maeda, Y. Shimodaira, B. Lee, K. Teramura, D. Lu, H. Kobayashi, K. Domen, *J. Phys. Chem. C* **111**, 18264 (2007).

553. K. Maeda, B. Lee, D. Lu, K. Domen, *Chem. Mater.* **21**, 2286 (2009).
554. J. Fang, F. Shi, J. Bu, J. Ding, S. Xu, J. Bao, Ma, Y.; Z. Jiang, W. Zhang, C. Gao, W. Huang, *J. Phys. Chem. C* **114**, 7940 (2010).
555. G. Liu, Y. Zhao, C. Sun, F. Li, G. Q. Lu, H. M. Cheng, Angew. *Chem. Int. Ed.* **47**, 4516 (2008).
556. Y. Li, G. Ma, S. Peng, G. Lu, S. Li, *Appl. Surf. Sci.* **254**, 6831 (2008).
557. R. Azouani, S. Tieng, K. Chhor, J. F. Bocquet, P. Eloy, E. M. Gaigneaux, K. Klementiev, A. V. Kanaev, *Phys. Chem. Chem. Phys.* **12**, 11325–11334 (2010).
558. W. J. Chun, A. Ishikawa, H. Fujisawa, T. Takata, J. N. Kondo, M. Hara, M. Kawai, Y. Matsumoto, K. Domen, *J. Phys. Chem. B* **107**, 1798 (2003).
559. G. Hitoki, T. Takata, J. N. Kondo, M. Hara, H. Kobayashi, K. Domen, *Chem. Commun.* **16**, 1698 (2002).
560. M. Hara, T. Takata, J. N. Kondo, K. Domen, *Catal. Today* **90**, 313 (2004).
561. G. Hitoki, A. Ishikawa, T. Takata, J. N. Kondo, M. Hara, K. Domen, *Chem. Lett.* **31**, 736 (2002).
562. R. Abe, T. Takata, H. Sugihara, K. Domen, *Chem. Commun.* **30**, 3829 (2005).
563. K. Maeda, N. Nishimura, K. Domen, *Appl. Catal. A: Gen.* **370**, 88 (2009).
564. J. Wang, S. Yin, K. Masakazu, Q. Zhang, S. Fumio, S. Tsugio, *J. Photochem. Photobiol. A: Chem.* **165**, 149 (2004).
565. J. Wang, H. Li, H. Li, S. Yin, T. Sato, *Solid State Sci.* **11**, 182 (2009).
566. A. Kasahara, K. Nukumizu, G. Hitoki, T. Takata, J. N. Kondo, M. Hara, H. Kobayashi, K. Domen, *J. Phys. Chem. A* **106**, 6750 (2002).
567. A. Kasahara, K. Nukumizu, T. Takata, J. N. Kondo, M. Hara, H. Kobayashi, K. Domen, *J. Phys. Chem. B* **107**, 791 (2003).
568. N. Nishimura, B. Raphael, K. Maeda, L. Le Gendre, R. Abe, J. Kubota, K. Domen, *Thin Solid Films*, **518**, 5855 (2010).
569. D. Yamasita, T. Takata, M. Hara, J. N. Kondo, K. Domen, *Solid State Ionics* **172**, 591 (2004).
570. T. Mishima, M. Matsuda, M. Miyake, *Appl. Catal. A: Gen.* **324**, 77 (2007).
571. K. R. Reyes-Gil, E. A. Reyes-Garca, D. Raftery, *J. Phys. Chem. C* **111**, 14579 (2007).
572. Y. Sun, C. J. Murphy, K. R. Reyes-Gil, E. A. Reyes-Garca, J. P. Lilly, D. Raftery, *Int. J. Hydrogen Energy* **33**, 5967 (2008).

573. S. Ge, H. Jia, H. Zhao, Z. Zheng, L. J. Zhang, *Mater. Chem.* **20**, 3052 (2010).
574. T. Hisatomi, K. Hasegawa, K. Teramura, T. Takata, M. Hara, K. Domen, *Chem. Lett.* **36**, 558 (2007).
575. K. M. Parida, S. Martha, D. P. Das, N. J. Biswal, *Mater. Chem.* **20**, 7144 (2010).
576. L. Jiang, Q. Wang, C. Li, J. Yuan, W. Shangguan, *Int. J. Hydrogen Energy* **35**, 7043 (2010).
577. A. Ishikawa, T. Takata, J. N. Kondo, M. Hara, H. Kobayashi, K. Domen, *J. Am. Chem. Soc.* **124**, 13547 (2002).
578. A. Ishikawa, Y. Yamada, T. Takata, J. N. Kondo, M. Hara, H. Kobayashi, K. Domen, *Chem. Mater.* **15**, 4442 (2003).
579. M. Yashima, K. Ogisub, K. Domen, *Acta Cryst.* **B64**, 291 (2008).
580. A. Ishikawa, T. Takata, T. Matsumura, J. N. Kondo, M. Hara, H. Kobayashi, K. Domen, *J. Phys. Chem. B* **108**, 2637 (2004).
581. K. Ogisu, A. Ishikawa, K. Teramura, K. Toda, M. Hara, K. Domen, *Chem. Lett.* **36**, 854 (2007).
582. K. Ogisu, A. Ishikawa, Y. Shimodaira, T. Takata, H. Kobayashi, K. Domen, *J. Phys. Chem. C* **112**, 11978 (2008).
583. K. Ikeue, S. Ando, T. Mitsuyama, Y. Ohta, K. Arayama, A. Tsutsumi, M. Machida, *Top. Catal.* **47**, 175 (2008).
584. K. Kobayakawa, A. Teranishi, T. Tsurumaki, Y. Sato, A. Fujishima, *Electrochim. Acta* **37**, 465 (1992).
585. L. Zheng, Y. Xu, Y. Song, C. Wu, M. Zhang, Y. Xie, *Inorg. Chem.* **48**, 4003 (2009).
586. W. J. Fan, Z. F. Zhou, W. B. Xu, Z. F. Shi, F. M. Ren, H. H. Ma, S. W. Huang, *Int. J. Hydrogen Energy* **35**, 6525 (2010).
587. S. Shen, L. Zhao, L. Guo, *Int. J. Hydrogen Energy* **35**, 10148–10154 (2010).
588. J. S. Jang, S. H. Choi, N. Shin, C. Yu, J. S. Lee, *J. Solid State Chem.* **180**, 1110 (2007).
589. D. Chen, J. Ye, *J. Phys. Chem. Solids* **68**, 2317 (2007).
590. M. Tabata, K. Maeda, T. Ishihara, Minegishi, T. Takata, K. Domen, *J. Phys. Chem. C* **114**, 11215 (2010).
591. D. Yokoyama, T. Minegishi, K. Maeda, M. Katayama, J. Kubota, A. Yamada, M. Konagai, K. Domen, *Electrochem. Commun.* **12**, 851 (2010).
592. A. Kudo, A. Nagane, I. Tsuji, H. Kato, *Chem. Lett.* **31**, 882 (2002).
593. J. Xu, Y. Ao, D. Fu, *Appl. Surf. Sci.* **256**, 884 (2009).
594. S. Song, J. Tu, L. Xu, X. Xu, Z. He, J. Qiu, J. Ni, J. Chen, *Chemosphere* **73**, 1401 (2008).

595. C. Liu, X. Tang, C. Mo, Z. Qiang, *J. Solid State Chem.* **181**, 913 (2008).
596. C. Wei, X. Tang, J. Liang, S. Tan, *J. Environ. Sci.* **19**, 90 (2007).
597. B. Tryba, *Int. J. Photoenergy* 721824 (2008). doi:10.1155/2008/721824
598. X. Z. Shen, J. Guo, Z. C. Liu, S. M. Xie, *Appl. Surf. Sci.* **254**, 4726 (2008).
599. K. Lv, H. Zuo, J. Sun, K. Deng, S. Liu, X. Li, D. Wang, *J. Hazard. Mater.* **161**, 396 (2009).
600. Y. Wang, Y. Wang, Y. Meng, H. Ding, Y. Shan, X. Zhao, X. Tang, *J. Phys. Chem. C* **112**, 6620 (2008).
601. Y. Huang, W. Ho, Z. Ai, X. Song, L. Zhang, S. Lee, *Appl. Catal. B: Environ.* **89**, 398 (2009).
602. X. Zhang, Q. Liu, *Appl. Surf. Sci.* **254**, 4780 (2008).
603. H. Xia, H. Zhuang, D. Xiao, T. Zhang, *J. Alloys Compd.* **465**, 328 (2008).
604. Z. He, X. Xu, S. Song, L. Xie, J. Tu, J. Chen, B. Yan, *J. Phys. Chem. C* **112**, 16431 (2008).
605. Z. Liu, Y. Zhou, Z. Li, Y. Wang, C. Ge, *Rare Met.* **26**, 263 (2007).
606. Xu, J.; Y. Ao, D. Fu, C. Yuan, *J. Colloid Interface Sci.* **328**, 447 (2008).
607. R. Long, N. English, *J. Chem. Phys. Lett.* **478**, 175 (2009).
608. J. Li, J. Xu, W. L. Dai, H. Li, K. Fan, *Appl. Catal. B: Environ.* **82**, 233 (2008).
609. Y. Shen, T. Xiong, H. Du, H. Jin, J. Shang, K. Yang, *J. Sol-Gel Sci. Technol.* **50**, 98 (2009).
610. L. H. Huang, C. Sun, Y. L. Liu, *Appl. Surf. Sci.* **253**, 7029 (2007).
611. M. Miyauchi, M. Takashio, H. Tobimatsu, *Langmuir* **20**, 232 (2004).
612. J. Wang, S. Yin, M. Komatsu, T. Sato, *J. Eur. Ceram. Soc.* **25**, 3207 (2005).
613. W. Wei, Y. Dai, M. Guo, L. Yu, B. Huang, *J. Phys. Chem. C* **113**, 15046 (2009).
614. Y. Gai, J. Li, S. S. Li, J. B. Xia, S. H. Wei, *Phys. Rev. Lett.* **102**, 036402 (2009).
615. W. J. Yin, H. W. Tang, S. H. Wei, M. M. Al-Jassim, J. Turner, Y. F. Yan, *Phys. Rev B* **82**, 045106 (2010).
616. R. Sasikala, A. R. Shirole, V. Sudarsan, C. Sudakar, R. Naik, R. Rao, S. R. Bharadwaj, *Appl. Catal. A: Gen.* **377**, 47 (2010).
617. S. Shet, K. S. Ahn, T. Deutsch, H. L. Wang, R. Nuggehalli, Y. F. Yan, J. Turner, M. Al-Jassim, *J. Power Sources* **195**, 5801 (2010).

618. I. Tsuji, A. Kudo, *J. Photochem. Photobiol. A: Chem.* **156**, 249 (2003).
619. Z. Lei, G. Ma, M. Liu, W. You, H. Yan, G. Wu, T. Takata, M. Hara, K. Domen, C. Li, *J. Catal.* **237**, 322 (2006).
620. S. Shionoya and W. M. Yen (eds.), *Phosphor Handbook* (CRC Press, Boca Raton, 1999), p. 255.
621. C. X. Xu and X. W. Sun, *Int. J. Nanotechnol.* **1**, 452 (2004).
622. J. E. Nause, *III–Vs Review* **12**, 28 (1999).
623. S. S. Mao, *Int. J. Nanotechnol.* **1**, 42 (2004).
624. M. W. Shin and R. J. Trew, *Electron. Lett.* **31**, 489 (1995).
625. F. Hamdani, A. E. Botchkarev, H. Tang, W. Kim, and H. Morkoc, *Appl. Phys. Lett.* **71**, 3111 (1997).
626. T. Detchprohm, K. Hiramatsu, H. Amano, and I. Akasaki, *ibid.* **61**, 2688 (1992).
627. R. D. Vispute, V. Talyansky, S. Choopun, R. P. Sharma, T. Venkatesan, M. He, X. Tang, J. B. Halpern, M. G. Spencer, Y. X. Li, L. G. Salamanca-Riba, A. A. Iliadis, and K. A. Jones, *Appl. Phys. Lett.* **73**, 348 (1998).
628. A. Ohtomo, M. Kawasaki, T. Koida, H. Koinuma, T. Sakurai, Y. Yoshida, M. Sumiya, S. Fuke, T. Yasuda, and Y. Segawa, *Mater. Sci. Forum* **264**, 1463 (1998).
629. P. Sharma, A. Gupta, K. V. Rao, F. J. Owens, R. Sharma, R. Ahuja, J. M. Osorio Guillen, B. Johansson, and G. A. Gehring, *Nat. Mater.* **2**, 673 (2003).
630. M. Hirano, T. Ito, *Mater. Res. Bull.* **43**, 2196 (2008).
631. G. Li, T. Kako, D. Wang, Z. Zou, J. Ye, *J. Solid State Chem.* **180**, 2845 (2007).
632. B. Muktha, G. Madras, T. N. G. Row, *J. Photochem. Photobiol. A: Chem.* **187**, 177 (2007).
633. Z. Zou, J. Ye, K. Sayama, H. Arakawa, *Chem. Phys. Lett.* **343**, 303 (2001).
634. Z. Zou, J. Ye, H. Arakawa, *Solid State Commun.* **119**, 471 (2001).
635. J. Luan, Z. Zou, M. Lu, Y. Chen, *Mater. Chem. Phys.* **98**, 434 (2006).
636. Z. G. Yi, J. H. Ye, *Appl. Phys. Lett.* **91**, 254108 (2007).
637. W. Yao, J. Ye, *J. Phys. Chem. B* **110**, 11188 (2006).
638. W. Yao, J. Ye, *Chem. Phys. Lett.* **450**, 370 (2008).
639. Z. Li, Y. Wang, J. Liu, G. Chen, Y. Li, C. Zhou, *Int. J. Hydrogen Energy* **34**, 147 (2009).
640. D. Wang, T. Kako, J. Ye, *J. Am. Chem. Soc.* **130**, 2724 (2008).
641. D. Wang, T. Kako, J. Ye, *J. Phys. Chem. C* **113**, 3785 (2009).
642. H. Liu, J. Yuan, W. Shangguan, Y. Teraoka, *J. Phys. Chem. C* **112**, 8521 (2008).

643. Q. Wang, H. Liu, L. Jiang, J. Yuan, W. Shangguan, *Catal. Lett.* **131**, 160 (2009).
644. F. Chevіré, F. Tessier, R. Marchand, *Eur. J. Inorg. Chem.* **6**, 1223 (2006).
645. W. Luo, Z. Li, X. Jiang, T. Yu, L. Liu, X. Chen, J. Ye, Z. Zou, *Phys. Chem. Chem. Phys.* **10**, 6717 (2008).
646. K. Maeda, T. Takata, M. Hara, N. Saito, Y. Inoue, H. Kobayashi, K. Domen, *J. Am. Chem. Soc.* **127**, 8286 (2005).
647. K. Maeda, K. Teramura, D. Lu, T. Takata, N. Saito, Y. Inoue, K. Domen, *Nature* **440**, 295 (2006).
648. M. Yashima, K. Maeda, K. Teramura, T. Takata, K. Domen, *Chem. Phys. Lett.* **416**, 225 (2005).
649. K. Maeda, K. Teramura, H. Masuda, T. Takata, N. Saito, Y. Inoue, K. Domen, *J. Phys. Chem. B* **110**, 13107 (2006).
650. T. Hirai, K. Maeda, M. Yoshida, J. Kubota, S. Ikeda, M. Matsumura, K. Domen, *J. Phys. Chem. C* **111**, 18853 (2007).
651. K. Maeda, K. Teramura, T. Takata, M. Hara, N. Saito, K. Toda, Y. Inoue, H. Kobayashi, K. Domen, *J. Phys. Chem. B* **109**, 20504 (2005).
652. K. Maeda, H. Hashiguchi, H. Masuda, R. Abe, K. Domen, *J. Phys. Chem. C* **112**, 3447 (2008).
653. X. Sun, K. Maeda, M. L. Faucheur, K. Teramura, K. Domen, *Appl. Catal. A: Gen.* **327**, 114 (2007).
654. K. Maeda, K. Domen, *Chem. Mater.* **22**, 612 (2010).
655. M. Yashima, H. Yamada, K. Maeda, K. Domen, *Chem. Commun.* **14**, 2379 (2010).
656. T. Hisatomi, K. Maeda, K. Takanabe, J. Kubota, K. Domen, *J. Phys. Chem. C* **113**, 21458 (2009).
657. K. Maeda, K. Teramura, K. Domen, *J. Catal.* **254**, 198 (2008).
658. K. Maeda, H. Masuda, K. Domen, *Catal. Today* **147**, 173 (2009).
659. Y. Lee, H. Terashima, Y. Shimodaira, K. Teramura, M. Hara, H. Kobayashi, K. Domen, M. Yashima, *J. Phys. Chem. C* **111**, 1042 (2007).
660. Y. Lee, K. Teramura, M. Hara, K. Domen, *Chem. Mater.* **19**, 2120 (2007).
661. X. Wang, K. Maeda, Y. Lee, K. Domen, *Chem. Phys. Lett.* **457**, 134 (2008).
662. F. Tessier, P. Maillard, Y. Lee, C. Bleugat, K. Domen, *J. Phys. Chem. C* **113**, 8526 (2009).
663. K. Kamata, K. Maeda, D. Lu, Y. Kako, K. Domen, *Chem. Phys. Lett.* **470**, 90 (2009).

664. M. R. Hoffmann, S. T. Martin, W. Choi and D. Bahnemann, *Chem. Rev.* **95**, 69 (1995).
665. D. Bockelmann, M. Lindner and D. Bahnemann, in *Fine Particles Science and Technology*, ed. E. Pelizzetti, Kluwer Academic Publishers, The Netherlands, 1996, p. 675.
666. M. I. Litter and J. A. Navio, *J. Photochem. Photobiol. A: Chem.* **98**, 171 (1996).
667. R. I. Bickley, J. S. Lees, R. J. D. Tilley, L. Palmisano and M. Schiavello, *J. Chem. Soc., Faraday Trans.* **88**, 377 (1992).
668. N. Tian, Z. Y. Zhou, S. G. Sun, Y. Ding, and Z. L. Wang, *Science* **316**, 732735 (2007).
669. O. Bikondoa et al. Direct visualization of defect-mediated dissociation of water on TiO_2 (110). *Nature Mater.* **5**, 189–192 (2006).
670. O. Dulub, et al. *Science* **317**, 1052–1056 (2007).
671. X. Q. Gong, A. Selloni, M. Batzill, and U. Diebold, *Nature Mater.* **5**, 665–670 (2006).
672. U. Diebold, The surface science of titanium dioxide. *Surf. Sci. Rep.* **48**, 53–229 (2003).
673. A. G. Thomas, et al. *Phys. Rev. B* **67**, 035110 (2003).
674. L. Kavan, M. Grätzel, S. E. Gilbert, C. Klemenz, and H. J. Scheel, *J. Am. Chem. Soc.* **118**, 6716–6723 (1996).
675. X. Q. Gong and A. Selloni, *J. Phys. Chem. B* **109**, 19560–19562 (2005).
676. M. Lazzeri, A. Vittadini, and A. Selloni, *Phys. Rev. B* **63**, 155409 (2001).
677. M. Lazzeri, and A. Selloni, *Phys. Rev. Lett.* **87**, 266105 (2001)
678. A. S. Barnard, and L. A. Curtiss, *Nano Lett.* **5**, 1261–1266 (2005).
679. X. Chen, S. S. Mao, *Chem. Rev.* **107**, 2891–2959 (2007).
680. F. Izumi, *Bull. Chem. Soc. Jpn* **51**, 1771–1776 (1978).
681. H. Berger, H. Tang, and F. Lévy, *J. Cryst. Growth* **130**, 108–112 (1993).
682. K. F. Zmbov, J. L. Margrave, *J. Phys. Chem.* **71**, 2893–2895 (1967).
683. H. G. Yang, C. H. Sun, S. Z. Qiao, J. Zou, G. Liu, S. C. Smith, H. M. Cheng and G. Q. Lu, *Nature* **453**, 638 (2008).
684. G. A. Somorjai and Y. G. Borodko, *Catal. Lett.* **76**, 1 (2001).
685. Z. Konya, V. F. Puntes, I. Kiricsi, J. Zhu, A. P. Alivisatos, G. A. Somorjai, *Catal. Lett.* **81**, 137–140 (2002).
686. C. A. Witham, W. Huang, C.-K. Tsung, J. N. Kuhn, G. A. Somorjai, and F. D. Toste, *Nature* Chemistry **2**, 36 (2010).

687. T. van Buuren, L. N. Dinh, L. L. Chase, W. J. Siekhaus, and L. J. Terminello, *Phys. Rev. Lett.* **80**, 3803 (1998).
688. Jinghua Guo, *Int. J. Nanotechnology* **1–2**, 193 (2004).
689. Jinghua Guo, pp. 259–291, in *Nanosystem Characterization Tools in the Life Sciences*, edited by Challa Kumar, Wiley-VCH Verlag GmbH and Co. KgaA, Weinheim (2006).
690. K. S. Hamad, R. Roth, J. Rockenberger, T. van Buuren, and A. P. Alivisatos, *Phys. Rev. Lett.* **83**, 3474 (1999).
691. Jean-Yves Raty, Giulia Galli, C. Bostedt, Tony van Buuren, and Louis J. Terminello, *Phys. Rev. Lett.* **90**, 37401 (2003).
692. F. Tao et al., *Science* **322**, 932 (2008).
693. T. Ma, Q. Fu, H. Y. Su et al., *Chem. Phys. Chem.* **7**, 1013 (2009).
694. K. J. Andersson, F. Calle-Vallejo, J. Rossmeisl et al., *J. Am. Chem. Soc.* **6**, 2404 (2009).
695. T. Feng, M. E. Grass, M. Y. W. Zhang, D. R. Bucher, S. Aloni, J. R. Renzas, Z. Liu, M. Salmeron, G. A. Somorjai, *J. Am. Chem. Soc.* **132**, 8697–8703 (2010).
696. V. K. LaMer, R. H. Dinergar, *J. Am. Chem. Soc.* **72**, 4847–48854 (1950).
697. C. B. Murray, C. R. Kagan, M. G. Bawendi, *Annu. Rev. Mater. Sci.* **30**, 545–610 (2000).
698. Y. Chen, E. Johnson, X. Peng, *J. Am. Chem. Soc.* **129**, 10937–10947 (2007).
699. H. Zheng; R. K. Smith; Y. W. Jun; C. Kisielowski; U. Dahmen; Alivisatos, A. P. *Science* **324**, 1309–1312 (2009).
700. P. V. Kamat, *Chem. Rev.* **93**, 267–300 (1993).
701. L. Brus, *Appl. Phys.* A **53**, 465 (1991).
702. H. Weller, *Adv. Mater.* **5**, 88 (1993).
703. H. Weller, A. Eychmuller, R. Vogel, L. Katsikas, A. Hasselbarth, M. Giersig, Isr. *J. Chem.* **33**, 107 (1993).
704. H. Weller, *Angew. Chem.* **32**, 41 (1993).
705. M. Grätzel, *Nature* **349**, 740 (1991).
706. A. Henglein, *Top. Curr. Chem.* **143**, 113 (1988).
707. M. L. Steigerwald, A. P. Alivisatos, J. M. Gibson, T. D. Harris, R. Kortan, A. J. Muller, A. M. Thayer, T. M. Duncan, D. C. Douglas, L. E. Brus, *J. Am. Chem. Soc.* **110**, 3046 (1988).
708. M. L. Steigerwald, L. E. Brus, L. E. *Annu. Rev. Mater. Sci.* **19**, 471 (1989).
709. M. G. Bawendi, M. L. Steigerwald, L. E. Brus, L. E. *Annu. Rev. Phys. Chem.* **41**, 477–496 (1990).
710. M. G. Bawendi, W. L. Wilson, L. Rothberg, P. J. Carroll, T. M. Jedju, M. Steigerwald, L. E. Brus, *Phys. Rev. Lett.* **65**, 1623 (1990).

711. N. G. Bawendi, P. J. Carroll, W. L. Wilson, L. E. Brus, *J. Chem. Phys.* **96**, 946 (1992).
712. R. A. Marcus, N. Sutin, *Biochim. Biophys. Acta* **811**, 265 (1985).
713. R. A. Marcus, *J. Phys. Chem.* **94**, 1050 (1990).
714. N. S. Lewis, *Annu. Rev. Phys.* **42**, 543 (1991).
715. A. J. Hoffman, H. Yee, G. Mills, M. R. Hoffmann, *J. Phys. Chem.* **96**. 5546 (1992).
716. A. J. Hoffman, H. Yee, G. Mills, M. R. Hoffmann, *J. Phys. Chem.* **96**, 5540 (1992).
717. M. Anpo, T. Shima, S. Kodama, Y. Kubokawa, *J. Phys. Chem.* **91**, 4305 (1987).
718. J. M. Nedelikovic, M. T. Nenadovic. O. I. Micic., K. A. J. Nozi, *J. Phys. Chem.* **90**, 12 (1986).
719. Y. Nosaka, N. Ohta, H. Miyama, *J. Phys. Chem.* **94**, 3752 (1990).
720. B. C. Faust, M. R. Hoffmann, D. W. Bahnemann, *J. Phys. Chem.* **93**, 6371 (1989).
721. W. Lee, Y.-M. Gao, K. Dwight, A. Wold, *Mater. Res. Bull.* **27**, 685 (1992).
722. M. R. Hoffmann, S. T. Martin, W. Choi, and D. W. Bahnemann, *Chem. Rev.* **95**, 69–96 (1995).
723. K. L. Frindell, M. H. Bartl, M. R. Robinson, G. C. Bazan, A. Popitsch, G. D. Stucky, *J. Solid State Chem.* **172**, 81 (2003).
724. Y. Bessekhouad, D. Robert, J. V. Weber, N. Chaoui, *J. Photochem. Photobiol.* A **167**, 49 (2004).
725. J. D. Bryan, S. M. Heald, S. A. Chambers, D. R. Gamelin, *J. Am. Chem. Soc.* **126**, 11640 (2004).
726. W. Choi, A. Termin, M. R. Hoffmann, *J. Phys. Chem.* **98**, 13669 (1994).
727. W. Choi, A. Termin, M. R. Hoffmann, *Angew. Chem.* **106**, 1148 (1994).
728. Y. J. Choi, A. Banerjee, A. Bandyopadhyay, S. Bose, *Ceram. Trans.* **159**, 67 (2005).
729. F. Coloma, F. Marquez, C. H. Rochester, J. A. Anderson, *Phys. Chem. Chem. Phys.* **2**, 5320 (2000).
730. F. Gracia, J. P. Holgado, A. Caballero, A. R. Gonzalez-Elipe, *J. Phys. Chem. B* **108**, 17466 (2004).
731. J. M. Herrmann, J. Disdier, P. Pichat, *Chem. Phys. Lett.* **108**, 618 (1984).
732. F. B. Li, X. Z. Li, M. F. Hou, *Appl. Catal. B* **48**, 185 (2004).
733. J. Li, S. Luo, W. Yao, Z. Zhang, *Mater. Lett.* **57**, 3748 (2003).
734. W. Li, Y. Wang, H. Lin, S. I. Shah, C. P. Huang, D. J. Doren, S. A. Rykov, J. G. Chen, M. A. Barteau, *Appl. Phys. Lett.* **83**, 4143 (2003).

735. W. Mu, J. M. Herrmann, P. Pichat, *Catal. Lett.* **3**, 73 (1989).
736. K. Nagaveni, M. S. Hegde, G. Madras, *J. Phys. Chem.* B **108**, 20204 (2004).
737. L. Palmisano, V. Augugliaro, A. Sclafani, M. Schiavello, *J. Phys. Chem.* **92**, 6710 (1988).
738. N. J. Peill, L. Bourne, M. R. Hoffmann, *J. Photochem. Photobiol.* A **108**, 221 (1997).
739. S. Peng, Y. Li, F. Jiang, G. Lu, S. Li, *Chem. Phys. Lett.* **398**, 235 (2004).
740. Y. Choi, T. Umebayashi, M. Yoshikawa, *J. Mater. Sci.* **39**, 1837 (2004).
741. M. Salmi, N. Tkachenko, R. J. Lamminmaeki, S. Karvinen, V. Vehmanen, H. Lemmetyinen, *J. Photochem. Photobiol.*, A **175**, 8 (2005).
742. J. Soria, J. C. Conesa, V. Augugliaro, L. Palmisano, M. Schiavello, A. Sclafani, *J. Phys. Chem.* **95**, 274 (1991).
743. A. Szabo, A. Urda, *Prog. Catal.* **11**, 73 (2002).
744. A. Szabo, A. Urda, *Prog. Catal.* **12**, 51 (2003).
745. C. Y. Wang, C. Boettcher, D. W. Bahnemann, J. K. Dohrmann, *Chem. Commun.* 1539 (2000).
746. C. Y. Wang, C. Boettcher, D. W. Bahnemann, J. K. Dohrmann, *J. Nanopart. Res.* **6**, 119 (2004).
747. W. Y. Wang, D. F. Zhang, X. L. Chen, *J. Mater. Sci.* **38**, 2049 (2003).
748. Y. Wang, Y. Hao, H. Cheng, J. Ma, W. Li, S. Cai, *J. Mater. Sci.* **34**, 3721 (1999).
749. Y. Wang, Y. Hao, H. Cheng, J. Ma, B. Xu, W. Li, S. Cai, *J. Mater. Sci.* **34**, 2773 (1999).
750. Y. Wang, H. Cheng, Y. Hao, J. Ma, B. Xu, W. Li, S. Cai, *J. Mater. Sci. Lett.* **18**, 127 (1999).
751. Y. Wang, H. Cheng, Y. Hao, J. Ma, W. Li, S. Cai, *Thin Solid Films* **349**, 120 (1999).
752. X. H. Xu, M. Wang, Y. Hou, W. F. Yao, D. Wang, H. Wang, *J. Mater. Sci. Lett.* **21**, 1655 (2002).
753. H. Yamashita, Y. Ichihashi, M. Takeuchi, S. Kishiguchi, M. Anpo, *J. Synchrotron Radiat.* **6**, 451 (1999).
754. X. H. Wang, J. G. Li, H. Kamiyama, M. Katada, N. Ohashi, Y. Moriyoshi, T. Ishigaki, *J. Am. Chem. Soc.* **127**, 10982 (2005).
755. W. Choi, A. Termin, M. R. Hoffmann, *J. Phys. Chem.* **98**, 13669–13679 (1994).
756. W. Choi, A. Termin, M. R. Hoffmann, *Angew. Chem.* **106**, 1148–1149 (1994).

757. W. Choi, A. Termin, M. R. Hoffmann, *Angew. Chem., Int. Ed.* Engl. **33**, 1091 (1994).
758. D. E. Scaife, Sol. Energy **25**, 41 (1980).
759. R. Abe, H. Takami, N. Murakami and B. Ohtani, *J. Am. Chem. Soc.* **130**, 7780 (2008).
760. J. Yu, J. Xiong, B. Cheng, Y. Yu and J. Wang, *J. Solid State Chem.* **178**, 1968 (2005).
761. J. Tang, Z. Zou and J. Ye, *Catal. Lett.* **92**, 53 (2004).
762. F. Amano, K. Nogami, R. Abe and B. Otani, *Chem. Lett.* **36**, 1314 (2007).
763. F. Amano, K. Nogami, R. Abe and B. Otani, *J. Phys. Chem.* C **112**, 9320 (2008).
764. H. Fu, C. Pan, W. Yao and Y. Zhu, *J. Phys. Chem.* B **109**, 22432 (2005).
765. C. Zhang and Y. Zhu, *Chem. Mater.* **17**, 3537 (2005).
766. S. Zhang, C. Zhang, Y. Man and Y. Zhu, *J. Solid State Chem.* **179**, 62 (2006).
767. S. Zhu, T. Xu, H. Fu, J. Zhao and Y. Zhu, *Environ. Sci. Technol.* **41**, 6234 (2007).
768. J. Wu, F. Duan, Y. Zheng and Y. Xie, *J. Phys. Chem.* C **111**, 12866 (2007).
769. J. Li, X. Zhang, Z. Ai, F. Jia, L. Zhang and J. Lin, *J. Phys. Chem.* C **111**, 6832 (2007).
770. S. Zhang, J. Shen, H. Fu, W. Dong, Z. Zheng and L. Shi, *J. Solid State Chem.* **180**, 1165 (2007).
771. H. Xie, D. Shen, X. Wang and G. Shen, *Mater. Chem. Phys.* **103**, 334 (2007).
772. H. Fu, C. Pan, L. Zhang and Y. Zhu, *Mater. Res. Bull.* **42**, 696 (2007).
773. L. Wu, J. Bi, Z. Li, X. Wang and X. Fu, *Catal. Today* **131**, 15 (2008).
774. H. Fu, S. Zhang, T. Xu, Y. Zhu and J. Chen, *Environ. Sci. Technol.* **42**, 2085 (2008).
775. L. Zhou, W. Wang and L. Zhang, *J. Mol. Catal. A: Chem.* **268**, 195 (2007).
776. C. A. Martinez-de la, S. O. Alfaro, E. L. Cuellar and U. O. Mendez, *Catal. Today* **129**, 194 (2007).
777. K. Tomita, V. Petrykin, M. Kobayashi, M. Shiro, M. Yoshimura and M. Kakihana, *Angew. Chem., Int. Ed.* **45**, 2378 (2006).
778. Jun Zhong, Jauwern Chiou, Chungli Dong, Li Song, Chang Liu, Sishen Xie, Huiming Cheng, Way-Faung Pong, Chinglin Chang, Yangyuan Chen, Ziyu Wu, Jinghua Guo, *Appl. Phys. Lett.* **93**, 023107 (2008).

779. J.-E. Rubensson, N. Wassdahl, G. Bray, J. Rindstedt, R. Nyholm, S. Cramm, N. Mårtensson, and J. Nordgren, *Phys. Rev. Lett.* **60**, 1759 (1987).
780. F.Kh. Gelímukhanov, L.N. Mazalov, and A.V. Kontratenko, *Chem. Phys. Lett.* **46**, 133 (1977).
781. Y. Ma, N. Wassdahl, P. Skytt, J.-H. Guo, J. Nordgren, J.E. Rubensson, T. Boske, W. Eberhardt, and S. Kevan, *Phys. Rev. Lett.* **69**, 2598 (1992).
782. F.Kh. Gelímukhanov and H. Ågren, *Phys. Rev.* A **49**, 4378 (1994).
783. F.Kh. Gelímukhanov, and H. Ågren, *Phys. Rev.* A **50**, 1129 (1994).
784. Y. Luo, H. Ågren, and F.Kh. Gelímukhanov, *J. Phys. B: At. Mol. Opt. Phys.* **27**, 4169 (1995).
785. Y. Luo, H. Ågren, F.Kh. Gelímukhanov, J.-H. Guo, P. Skytt, N. Wassdahl, and J. Nordgren, *Phys. Rev.* B **52**, 14479 (1995).
786. S.M. Butorin, J.-H. Guo, M. Magnuson, P. Kuiper, and J. Nordgren, *Phys. Rev.* B **54**, 4405 (1996).
787. P. Kuiper, J.-H. Guo, C. Såthe, L.-C. Duda, J. Nordgren, J.J.M. Pothuizen, F.M.F. de Groot, and G.A. Sawatzky, *Phys. Rev. Lett.* **80**, 5204 (1998).
788. Jinghua Guo, *Int. J. Nanotechnology* **1–2**, 193 (2004).
789. Jinghua Guo, pp. 259–291, in *Nanosystem Characterization Tools in the Life Sciences*, edited by Challa Kumar, Wiley-VCH Verlag GmbH & Co. KgaA, Weinheim (2006).
790. Jun Zhong, Jauwern Chiou, Chungli Dong, Li Song, Chang Liu, Sishen Xie, Huiming Cheng, Way-Faung Pong, Chinglin Chang, Yangyuan Chen, Ziyu Wu, Jinghua Guo, *Appl. Phys. Lett.* **93**, 023107 (2008).
791. B. O'Regan and M. Gratzel, *Nature* **353**, 737 (1991).
792. A. Hagfeldt, N. Vlachopoulos, and M. Gratzel, *J. Electrochem. Soc.* **141**, L82 (1994).
793. S. Y. Huang, L. Kavan, I. Exnar, and M. Gratzel, *J. Electrochem. Soc.* **142**, L142 (1995).
794. L. Vayssieres, C. Sthe, S. M. Butorin, D. K. Shuh, J. Nordgren, and J.-H. Guo, *Adv. Mater.* **17**, 2320 (2005).
795. V. F. Puntes, K. M. Krishnan, A. P. Alivisatos, *Science* **291**, 2115–2117 (2001).
796. V. F. Puntes, P. Gorostiza, D. M. Aruguete, N. G. Bastus, A. P. Alivisatos, *Nat. Mater.* **3**, 263–268 (2004).
797. H. Liu, J.-H. Guo, Y. Yin, A. Augustsson, C. Dong, J. Nordgren, C. Chang, P. Alivisatos, G. Thornton, D. F. Ogletree, F. G.

Requejo, F. de Groot, and M. Salmeron, *Nano Lett.* **7**, 1919 (2007).
798. M. E. Dry, *Catalysis: Science and Technologies;* Editors: J. R. Anderson and M. Boudart, Springer-Verlag: Berlin, Vol. 1 p 159 (1981).
799. A. Y. Khodakov, W. Chu, P. Fongarland, *Chem. Rev.* **107**, 1692 (2007).
800. T. Herranz, X. Deng, A. Cabot, J.-H. Guo, and M. Salmeron, *J. Phys. Chem.* B **113**, 10721–10727 (2009).
801. F. Jiao and H. Frei, *Angew. Chem.* **121**, 1873 (2009).
802. J.-H. Guo, L. Vayssieres, C. Persson, R. Ahuja, B. Johansson, and J. Nordgren, *J. Phys.: Condens. Matter* **14**, 6969 (2002).
803. C. L. Dong, C. Persson, L. Vayssieres, A. Augustsson, T. Schmitt, M. Mattesini, R. Ahuja, C. L. Chang, and J.-H. Guo, *Phys. Rev.* B **70**, 195325 (2004).
804. M. Huang, S. Mao, H. Feick, H. Yan, Y. Wu, H. Kind, E. Weber, R. Russo, and P. Yang, *Science* **292**, 1897 (2001).
805. Y. Marcus, *Journal of Physical Chemistry* **91**, 4422–4428 (1987).
806. A. K. Soper, G. W. Neilson, J. E. Enderby, and R. A. Howe, *Physics* **10**, 1793–1801 (1977).
807. G. W. Neilson, J. E. Enderby, *Journal of Physics C-Solid State Physics* **11**, L625–L628 (1978).
808. K. Waizumi, T. Kouda, A. Tanio, N. Fukushima, and H. Ohtaki, *Journal of Solution Chemistry* **28**, 83–100 (1999).
809. M. Magini, *Journal of Chemical Physics* **74**, 2523–2529 (1981).
810. R. Caminiti and P. Cucca, *Chemical Physics* Letters **89**, 110–114 (1982).
811. F. de Groot, Multiplet effects in x-ray spectroscopy, *Coordination Chemistry Reviews* **249**, 31–63 (2005).
812. G. Vanderlaan, J. Zaanen, G. A. Sawatzky, R. Karnatak, and J. M. Esteva, *Physical Review* B **33**, 4253–4263 (1986).
813. M. Armand, J.M. Tarascon, *Nature* **451**, 652 (2008).
814. M. Lefevre, E. Proietti, F. Jaouen, J.P. Dodelet, *Science* **324**, 71 (2009).
815. F.U. Renner, A. Stierle, H. Dosch, D.M. Kolb, T.L. Lee, J. Zegenhagen, *Nature* **439**, 707 (2006).
816. J.A. Switzer, H.M. Kothari, P. Poizot, S. Nakanishi, E.W. Bohannan, *Nature* **425**, 490 (2003).
817. P.G. Slade, *Electrical Contacts: Principles and Applications*, Marcel Dekker, New York, 1999.
818. A.J. Bard, L.R. Faulkner, *Electrochemical Methods: Fundamentals and Applications*, Wiley, 2001.

819. M.J. Weaver, *J. Phys. Chem.* **100**, 13079 (1996).
820. P. Jiang, J.-L. Chen, F. Borondics, P.-A. Glans, M. W. West, C.-L. Chang, M. Salmeron, J.-H. Guo, *Electrochemistry Communications* **12**, 820 (2010).
821. M.P. Sánchez, M. Barrera, S. González, R.M. Souto, R.C. Salvarezza, A.J. Arvia, *Electrochim. Acta* **35**, 1337 (1990).
822. M. Drogowska, L. Brossard, H. Ménard, *J. Electrochem. Soc.* **139**, 39 (1992).
823. K.O. Kvashnina, S.M. Butorin, A. Modin, I. Soroka, M. Marcellini, J.H. Guo, L. Werme, J. Nordgren, *J. Phys.: Condens. Matter* **19**, 226002 (2007).
824. M. Magnuson, N. Wassdahl, J. Nordgren, *Phys. Rev. B* **56**, 12238 (1997).
825. M. Hävecker, A. Knop-Gericke, T. Schedel-Niedrig, R. Schlögl, *Angew. Chem. Int. Ed.* **37**, 1939 (1998).

Index

T

V

W

X

Z